To my parents and brother who encouraged
all my crazy hobbies.
To my wife who now has to deal with the result.

Contents

Welcome!

By opening up this book, you have taken the first step in joining the world of amateur radio! Congratulations! Amateur radio has something to offer to everyone! There are always new things to do and try no matter how long you've had a license. With amateur radio, the whole world is available to you through a little transceiver in your workshop. You have made a great choice!

What is Amateur Radio All About?

Amateur radio, or ham radio, has existed since the earliest days of wireless communication, back when if you owned wireless set, you most likely built it yourself. This engineering spirit of was the source of some of the greatest advances in wireless technology to which we all owe a great deal of thanks. Hams are driven by a spirit of community, goodwill, and communication. They come from all ages and backgrounds. They are all over the world! You do not need to be an electrical engineer or a physics genius to become an amateur radio operator. You just need the desire to become one!

In the United States, to become a licensed amateur operator you must pass a test following the guidelines dictated by the Federal Communication Commission (FCC). There are three class of licenses: Technician, General, and Extra. The Technician class is the initial license all amateurs now receive. As you continue through various license classes, your amateur privileges increase which gives you an incentive to upgrade your license. Those Technician licensees who wish to expand their ham radio experience, may take the General class exam to get on the shortwave bands and increased access to worldwide communications. Those General class operators can expand their frequency privileges even more by taking the Extra class examination. It is all up to you!

What you can do with amateur radio is only limited to your imagination. Hams are developing new ways to communicate as well as creating new activities all the time. There are modes of communication and ham events out there for every taste and personality. You can try your hand at transmitter hunting, storm chasing, contesting, low power operation, "home-brewing" (making your own equipment), computer and data communication, bouncing signals off the moon or meteor showers, satellite communication, talking to astronauts on the International Space Station, or you can just sit back and tune through the frequencies looking for another ham to chat with. Whatever your tastes, amateur radio has something for you!

Along with all the fun stuff, amateurs provide a huge service to their communities. Whenever disaster strikes or in times of national emergency the world looks to ham radio operators to provide a vital source of communications. When the internet, telephones, and cell phones are down, there's nothing but amateur radio! The fun we get to have by talking to one another is just the privilege we are allowed to develop our radio skills for when they are truly needed. In the United States, the amateur radio service is made up entirely of volunteer amateurs. You are strictly forbidden to make money in the use of you FCC granted amateur privileges. Amateur radio is strictly a volunteer service of communication experts who provide a service to the community. That's the real reason the amateur service exists and one of the responsibilities you are taking on for your community. You are a patriot!

Amateur radio is what you make of it. There is something in it for everyone! It is one of the most diverse hobbies out there and you can do as much or as little as you want with it. Some hams are content to just talk on the 2 meter repeater while commuting to and from work. Others spend hours designing and building their own equipment. Whatever your tastes, you made a great decision in taking the first steps to get licensed!

What Can You Do With a Technician Class License?

The truth is an awful lot. Technician class operators make up the majority of U.S. hams...so you have many people to talk to. The technician class operating privileges authorized by the FCC allow technician operators all the very high frequency (VHF) and ultra high frequency (UHF) amateur bands. What this means is a technician class operator can operate the same VHF and UHF frequencies that the general and extra class operators get. Since these are the more popular bands, there is always plenty to do!

The way the FCC breaks down the privileges for the various amateur license classes is by frequency and mode of communication. Only certain types of communication are allowed on certain bands. There are several different modes of communication. On all the VHF and UHF bands you can use radio-teletype (RTTY), data modes, phone, and image modes (Yes, you can send pictures and television images over the amateur bands!). We'll get into the specifics surrounding these modes a little later.

The VHF and UHF bands have a much smaller range than the high frequency (HF) bands. The HF bands are the ones that are most often used to get those long range international contacts. Hams call these type of contacts DX. The vast majority of the HF band privileges go to the general and extra class operators. One more reason to upgrade your license! Fear not! Technician operators do have voice privileges on the 10 meter band which can allow you to talk almost anywhere in the world when conditions are right. Also, the 6 meter band in the lower VHF range has been known for some incredible contacts as well. If you are willing to learn Morse code (otherwise known as CW in amateur circles) there is a segment on both the 40 meter and 80 meter bands which will allow you to get some of those international DX contacts! Once upon a time, there used to be an additional Morse code test you had to take to get your license, but now it is no longer required so you can breathe a sigh of relief. The complete list of technician class privileges are outlined in Figure 1.

Figure (1) Technician Class Privileges

Band	Frequency Range	Authorized Modes
80 meters	3.525 MHz to 3.600 MHz	CW only (limited to 200 watts)
40 meters	7.025 MHz to 7.125 MHz	CW only (limited to 200 watts)
15 meters	21.025 MHz to 21.200 MHz	CW only (limited to 200 watts)
10 meters	28 MHz to 28.3 MHz	RTTY and Data (limited to 200 watts)
	28.3 MHz to 28.5 MHz	SSB Phone (limited to 200 watts)
6 meters	50.0 MHz to 50.1 MHz	CW only
	50.1 MHz to 54.0 MHz	RTTY, Data, Phone, and Image
2 meters	144.0 MHz to 144.1 MHz	CW only
	144.1 MHz to 148.0 MHz	RTTY, Data, Phone, and Image
1.25 meters	219.0 MHz to 220.0 MHz	Fixed Digital Message Forwarding System
	222.0 MHz to 225.0 MHz	RTTY, Data, Phone, and Image
70 cm	420.0 MHz to 450.0 MHz	RTTY, Data, Phone, and Image
33 cm	902.0 MHz to 928.0 MHz	RTTY, Data, Phone, and Image
23 cm	1240 MHz to 1300 MHz	RTTY, Data, Phone, and Image

Despite the majority of technician class privileges being limited to the more local VHF and UHF type communications, there is still a lot going on. You can try your hand at satellite operation, bouncing signals off of the moon, packet radio (a popular form of data communication), emergency communications, or just sit back and chat on you local ham radio club's repeater. Don't get intimidated! If all of these acronyms and "ham speak" are throwing you off a bit, don't worry. You will be talking like a local by the end of this book! There is much fun to be had and you've already taken the first step!

The Exam

You enter the room. There are strange people everywhere. The guy in the corner looks like a rocket scientist (and probably is)! You take your seat. The girl on you right has a calculator that looks like it could calculate the human genome with just a few key punches. You start to sweat. You realize you're sweating. The girl with the calculator looks at you. You think she notices that you're sweating. You sweat more. You sit back, take a deep breath, look down and realize your wearing nothing but boxer shorts and a pair of black socks! You scream...then suddenly become aware that you are lying in your bed! It was all a dream.

The technician exam is nothing like this. Everyone gets a little test anxiety, but only when they are not prepared. You will be ready, Grasshopper! Trust me, there is no shock and awe associated with the technician exam, or any of the amateur exams for that matter. To make it even easier, I'm going to break down exactly what the exam consists of, what to bring, and what to expect.

The Technician Class exam consists of 35 questions taken from a pool of 396 questions published by the National Conference of Volunteer Examiner Coordinators' (NCVEC) Question Pool Committee. To pass, you need to score a 74% which means you need to answer at least 26 questions correctly. No sweat! The question pool is broken down into 10 subelements, T1 through T0. Each subelement covers a general amateur radio subject. For instance subelement T1 deals primarily with FCC rules and definitions. The subelements are further broken down into various numbers of sections which cover more specific topics within the subelement such as basic algebra level math questions, questions on electronics and electricity, rules and regulation, and propagation. There are thirty-five sections for the entire question pool. Thirty-five sections, thirty-five questions on the exam. Coincidence?

When you have completed your studies and are ready to take the exam, you face your next hurdle: finding a place to take the exam. If you are familiar with your local amateur radio club, or have a friend who is encouraging you to become a ham, they are probably your best resources in finding an exam session. If you are more on your own, there are plenty of resources to help you out.

Hamfests will often have an exam session as part of the day's activities. Hamfests are gatherings of radio amateurs and radio enthusiasts at which much trading and buying of equipment occur. They are usually sponsored by a local ham club and are a lot of fun. Besides the ability to buy and sell equipment at hamfests, the sponsors will often throw in some exam sessions and classes on the latest trends in amateur radio. There are usually specific times posted for exam sessions and most sessions at hamfests encourage walk-ins. This does not mean you do not have to pre-register if you have the opportunity to register to take the exam. Even though many encourage walk-ins, there is a limited amount of seats and time available. Register if you can and show up early.

There are a ton of online resources available to help you find a local exam session. The American Radio Relay League (ARRL) is a fabulous resource for all hams and has a great deal of material devoted to helping you prepare for and take the technician exam! If you check out the ARRL website, you can find a local exam session based on your zip code (www.arrl.org/find-an-amateur-radio-exam-session). Usually they will have a point of contact for the amateur operator coordinating the exam. These exam sessions vary in size and number of exams they can give. Also, there may be restrictions on what class of exam they are giving that day. I highly recommend contacting the exam session coordinator and let them know you are coming. You don't want to get all pumped up for the exam just find out they are not taking walk-ins!

The exam is administered by three Volunteer Examiners or VEs. These VEs are registered with one of the Volunteer Exam Coordinators, or VECs, which have special authority from the FCC to administer and process license exam sessions. Along with many other helpful roles supporting amateur radio, the ARRL also serves as a VEC. During the exam, these are the only three individuals you are allowed to talk to and what they say goes. The VEs will administer, collect, and grade the exams as well as process the necessary paperwork when you pass. Notice I said, "when you pass." If you suffer from an impairment which prevents you from taking a written exam, the VEs will accommodate you as much as reasonably possible. For instance, if you have trouble seeing, the Volunteer Examiners can give you the exam orally. There really is no excuse for not taking the exam!

There are a few things you will need to take with you to the exam. They are:

1. A photo ID. If you do not have a photo ID, you will need to bring two other forms of ID. Possible choices are a social security card, birth certificate, library card, or a bank or utility statement with your permanent address on it.

2. Your social security number. The VE will need to register your social security number with the FCC once you pass the exam.

3. A form 605. This is the form the Volunteer Examiners will send to the FCC once you pass the exam. You can find and print one from the FCC's document page at www.fcc.gov/formpage.html. Most exam sessions will have extra copies of the form 605, but it never hurts to be safe.

4. Two #2 pencils with erasers and a few sheets of scrap paper.

5. A calculator. You may need to show the VE the calculator's memory is erased before taking the exam.

6. A check or money order for the exam fees. If the VEC administering the exam is the ARRL, the fee for the exam will be $15. This is to cover the out of pocket expenses of the Volunteer Examiners. Keep in mind that these examiners are amateurs who are volunteering their time to give you the exam. Depending on the number of potential hams taking the exam, those Volunteer Examiners may be spending a fairly decent amount of money to set up and administer the examination. However, not all exam sessions require a fee for the examination. You should check with the person coordinating the exam to see if a fee is going to be charged. Beware! It is possible that if you decide to take more than one exam at the session you may need to pay an additional exam fee for the second exam. Another good question to ask.

Don't worry about the exam! Nobody wants you to fail! Everyone in the room is either a ham who is hoping you pass and join the amateur radio community or someone who is in the same position as you. It is not a competition. The guy who scores a 100% and the guy who scores a 74% are both getting call signs!

How This Book is Set Up

This book takes a four prong approach to learning the material: association of the correct answer with the each question, explanation, short term memory recall, and long term memory recall. The course material is divided up into 35 lessons; one for each section in the exam question pool. These are the same questions you will see on the exam. For each question, the correct answer is given alone and without the other incorrect multiple choice answers. This allows you to better associate the correct answer with the question without mucking your memory up with the incorrect answers. There's no point in studying the wrong answers! An explanation is then given on why it is the correct answer, any theory behind it, and any tricks to help make the correct answer stand out from the other possible answers on the exam.

At the end of each lesson there is a quiz consisting of the exact questions and multiple choice answers you may see on the exam from that section. Keep in mind that you may see only one question from each section on the actual exam. I strongly recommend you give the quizzes an honest shot. They are short and really help you cement the material in your brain. If you struggle with a question or series of questions, go back, review the material for that lesson, and take the whole quiz again. The goal is to get 100% on each quiz before you move on to the next lesson. Take your time, be patient, and don't expect to ace everything on the first try!

Unfortunately, some of the questions can only be learned by brute force memorization. It hurts just thinking about it. There are a few questions which ask stuff like what frequencies fall within which band, what formula do you need, how many volts can kill you, and stuff like that. For questions like these, there is a Study Sheet at the end of each lesson to help you focus your studies. There is no frivolous information on the Study Sheets, just the more obscure information you need to know for the exam. No reason to work harder than you need to. These sheets will come in very handy in your final preparation for the exam.

Once you have completed the 35 lessons, there are three practice exams to help build confidence and focus your review efforts before taking the actual exam. Each practice exam consists of 35 questions from the question pool, just like you would see them on the actual exam. These practice exams consist of three difficulty levels: Simple, Challenging, and Hard. The Simple Exam consists of the easier questions from each section. This helps get your mind back in gear to help recall some of the earlier lessons. The Challenging Exam is made up of those questions which require a little bit of thought and calculation to get correct. The Hard Exam is exactly that, hard. It consists of the most difficult questions from the question pool. The purpose behind these three exams is to review the major subject material of the course and allow you to identify areas where you might need some extra review before the big day. Try to avoid the temptation of going straight to the Hard Exam just to see how you would do. Taking the two easier exams first allows your brain to stretch out and better ease into the hard stuff. Once you feel comfortable with all three practice exams, you are ready!

This course is designed for the reader to complete one lesson per day. Each lesson will take about 15 to 20 minutes. If you are feeling confident and want to do more lessons per day, I do not recommend doing more than two and you really should take a several hour break between each lesson. Your brain needs time to process the information and too much at once can cause brain-fry. However, this is only my recommendation. Only you know the best ways for you to learn.

There are some things you shouldn't expect from this course. This course is meant to get you past the exam so you can safely, legally, and courteously start your ham radio experience. This course is not meant to make you the "human encyclopedia" of amateur radio knowledge. The purpose of these lessons is to coach to the exam. As most amateurs will agree, ham radio is a hands on hobby. You learn a lot more by doing and experiencing the hobby than you will by preparing for an exam. The purpose of the exam is to ensure the FCC that you have the ability and knowledge to avoid causing an international communications incident, violate federal laws, or hurt yourself and/or others. It is important stuff to know, but really only touches the tip of the amateur radio iceberg. Ham radio is not a "closed book" hobby. In fact, it's quite the opposite. Nobody is going to stop you from looking up authorized frequencies or operating procedures after you get your license. There is a ton of books and online information out there, not to mention an entire world of amateur radio operators that are willing to help you make the most of this great hobby!

The Questions

Before we get started with the lessons, it is important to review exactly how the questions are numbered. Once you take a look at the questions you'll notice that they are not numbered in the normal 1, 2, 3, 4..., fashion. Each question is assigned a unique code, if you will, that identifies which subelement and section it was taken from in the exam question pool. Though this may seem like a nightmare, this unique numbering system benefits you in studying for the exam. Let's take a look.

Take this question as an example:

T1B01 [97.3(a)(28)] What is the ITU?

Now let's separate out the question number and reference:

T1B01 [97.3(a)(28)]

The question number breaks down fairly easily. The "T1" tells you that this question comes from T1 Subelement of the question pool. The "B" lets you know that it is taken from Section B of that subelement. Put those two together and you can identify this question comes from the T1B section of the question pool. The "01" means that it is the first question in that section. The "[97.3(a)(28)]" is the reference to the question's answer. In this case, the question is referencing the FCC Part 97 rules which govern amateur radio (more on that later).

The way this works to your benefit is if you are struggling with a question on one of the quizzes or practice exams, this numbering system will tell you which lesson and question number to review. For this question, you would want to go back to review the first question in the T1B section found in Lesson 2 of the course.

Lesson 1
The T1A Section
The Amateur Radio Service

T1A01 [97.3(a)(4)] For whom is the Amateur Radio Service intended?
Answer: *Persons who are interested in radio technique solely with a personal aim and without pecuniary interest.*

The whole definition from the Part 97 rules goes:

A radio communication service for the purpose of self-training, intercommunication and technical investigations carried out by amateurs, that is, duly authorized persons interested in radio solely with a personal aim and without pecuniary interest. [FCC Part 97 Rules, 97.3(a)(4)]

This is you! You are a person who is interested in radio technique. Amateur radio is a hobby. No corporation or government is forcing you to be a ham. This makes amateur radio a personal aim. You are not seeking to use your amateur privileges to make money. It is solely a voluntary service. You're looking for the big picture in this answer. It's a broad definition for a broad hobby.

T1A02 [97.1] What agency regulates and enforces the rules for the Amateur Radio Service in the United States?
Answer: *The FCC.*

FCC stands for Federal Communications Commission. They are the government agency which oversee all communications activities within the United States. The FCC governs the airwaves. They make all the rules not only for amateur radio, but commercial and civil defense communications as well. They are also the agency which issues you your license!

T1A03 Which part of the FCC rules contains the rules and regulations governing the Amateur Radio Service?
Answer: *Part 97.*

You will probably know this one by heart by the time you finish this course, but in the mean time you might want to memorize it. The Part 97 rules is the governing document for amateur radio in the United States. It is very detailed. The Part 97 rules get updated often to accommodate new technologies and techniques in amateur radio. The FCC writes them and we stick to them!

T1A04 [97.3(a)(23)] Which of the following meets the FCC definition of harmful interference?
Answer: *That which seriously degrades, obstructs, or repeatedly interrupts a radio communication service operating in accordance with the Radio Regulation.*

So there you are, passing a message about damage from a recent flood to net control, when all the sudden a loud overpowering noise from another station cuts off your transmission! This is a type of harmful interference. You are looking for many negative words to separate this question out. Harmful interference is bad and degrades, obstructs, and/or interrupts. The other important piece of this answer is that the interference negatively disrupts any communications service that is operating in accordance with radio regulations. In other words, harmful interference interferes with a signal from a station which is obeying the rules.

T1A05 [97.3(a)(40)] What is the FCC Part 97 definition of a space station?
Answer: *An amateur station located more than 50 km above the Earth's surface.*

Yes, you can communicate with stations in space! The FCC definition for a space station deals with height above the Earth, not what kind of vehicle or craft is carrying the station. For this question, you want to aim high. Fifty kilometers is the highest altitude of the possible answers. A space station is an amateur station that is at least 50 km above the Earth's surface.

T1A06 [97.3(a)(43)] What is the FCC Part 97 definition of telecommand?
Answer: *A one-way transmission to initiate, modify or terminate functions of a device at a distance.*

A telecommand is any transmission that you send to control a device. A better way of putting it, it is a transmission in which you are sending a command. A telecommand is also a one-way transmission. Like any command, a telecommand is an order. You do not need to wait for a response to tell a device to do something. A telecommand is a one-way transmission to initiate, modify or terminate functions of a device at a distance.

T1A07 [97.3(a)(45)] What is the FCC Part 97 definition of telemetry?
Answer: *A one-way transmission of measurements at a distance from the measuring instrument.*

Telemetry is the same basic principle as a telecommand. In the case of telemetry, you are receiving a one-way transmission from a device at a distance. Telemetry transmissions provide you some sort of measurement or other similar information. So, telemetry is a one-way transmission of measurements at a distance from the measuring instrument.

T1A08 [97.3(a)(22)] Which of the following entities recommends transmit/receive channels and other parameters for auxiliary and repeater stations?
Answer: *Frequency Coordinator.*

A Frequency Coordinator is a person or group who ensures auxiliary and repeater stations all transmit in harmony without interfering with one another. Frequency coordinators are not appointed by the FCC. A Frequency Coordinator is selected by local or regional amateurs whose stations are eligible to be auxiliary or repeater stations. These types of stations usually operate on fixed frequencies and are frequently used. Working with a Frequency Coordinator helps ensure the most efficient use of the airwaves and keeps these stations from interfering with one another.

T1A09 [97.3(a)(22)] Who selects a Frequency Coordinator?
Answer: *Amateur operators in a local or regional area whose stations are eligible to be auxiliary or repeater stations.*

This should sound a little familiar. A Frequency Coordinator is selected by local or regional hams whose stations are eligible to be auxiliary or repeater stations. Again, a Frequency Coordinator is not appointed by the FCC or any other communications agency, for that matter. A Frequency Coordinator is selected its local or regional amateur radio operators.

T1A10 [97.3(a)(5)] What is the FCC Part 97 definition of an amateur station?
Answer: *A station in an Amateur Radio Service consisting of the apparatus necessary for carrying on radio communications.*

This is a fairly simple one. An amateur station is any functioning station in the amateur radio service. This means that the station is operated by a licensed amateur operator within the parameters outlined in the Part 97 rules. It is not the building where the station is located or just anyone who is using a radio and not getting paid. It is a station in an amateur radio service consisting of the apparatus necessary for carrying on radio communications.

T1A11 [97.3(a)(7)] Which of the following stations transmits signals over the air from a remote receive site to a repeater for retransmission?
Answer: *Auxiliary station.*

Some of the other possible answers can be easily mistaken to be correct for this question. To help make things make a little more sense, the FCC Part 97 definition of an auxiliary station describes it as, "An amateur station, other than a message forwarding system, that is transmitting communications point-to-point within a system of cooperating amateur stations." Auxiliary stations help extend the range of repeaters. A receiver at a remote site can receive a signal which an auxiliary station will then send to a repeater for retransmission.

Quiz #1

(Find the answers on page 256)

1. _____ T1A01 [97.3(a)(4)] For whom is the Amateur Radio Service intended?
A. Persons who have messages to broadcast to the public
B. Persons who need communications for the activities of their immediate family members, relatives and friends
C. Persons who need two-way communications for personal reasons
D. Persons who are interested in radio technique solely with a personal aim and without pecuniary interest

2. _____ T1A02 [97.1] What agency regulates and enforces the rules for the Amateur Radio Service in the United States?
A. FEMA
B. The ITU
C. The FCC
D. Homeland Security

3. _____ T1A03 Which part of the FCC rules contains the rules and regulations governing the Amateur Radio Service?
A. Part 73
B. Part 95
C. Part 90
D. Part 97

4. _____ T1A04 [97.3(a)(23)] Which of the following meets the FCC definition of harmful interference?
A. Radio transmissions that annoy users of a repeater
B. Unwanted radio transmissions that cause costly harm to radio station apparatus
C. That which seriously degrades, obstructs, or repeatedly interrupts a radio communication service operating in accordance with the Radio Regulations
D. Static from lightning storms

5. _____ T1A05 [97.3(a)(40)] What is the FCC Part 97 definition of a space station?
A. Any multi-stage satellite
B. An Earth satellite that carries one of more amateur operators
C. An amateur station located less than 25 km above the Earth's surface
D. An amateur station located more than 50 km above the Earth's surface

6. _____ T1A06 [97.3(a)(43)] What is the FCC Part 97 definition of telecommand?
A. An instruction bulletin issued by the FCC
B. A one-way radio transmission of measurements at a distance from the measuring instrument
C. A one-way transmission to initiate, modify or terminate functions of a device at a distance
D. An instruction from a VEC

7. _____ T1A07 [97.3(a)(45)] What is the FCC Part 97 definition of telemetry?
A. An information bulletin issued by the FCC
B. A one-way transmission to initiate, modify or terminate functions of a device at a distance
C. A one-way transmission of measurements at a distance from the measuring instrument
D. An information bulletin from a VEC

8. _____ T1A08 [97.3(a)(22)] Which of the following entities recommends transmit/receive channels and other parameters for auxiliary and repeater stations?
A. Frequency Spectrum Manager
B. Frequency Coordinator
C. FCC Regional Field Office
D. International Telecommunications Union

9. _____ T1A09 [97.3(a)(22)] Who selects a Frequency Coordinator?
A. The FCC Office of Spectrum Management and Coordination Policy
B. The local chapter of the Office of National Council of Independent Frequency Coordinators
C. Amateur operators in a local or regional area whose stations are eligible to be auxiliary or repeater stations
D. FCC Regional Field Office

10. _____ T1A10 [97.3(a)(5)] What is the FCC Part 97 definition of an amateur station?
A. A station in an Amateur Radio Service consisting of the apparatus necessary for carrying on radio communications
B. A building where Amateur Radio receivers, transmitters, and RF power amplifiers are installed
C. Any radio station operated by a non-professional
D. Any radio station for hobby use

11. _____ T1A11 [97.3(a)(7)] Which of the following stations transmits signals over the air from a remote receive site to a repeater for retransmission?
A. Beacon station
B. Relay station
C. Auxiliary station
D. Message forwarding station

Lesson 1 Study Sheet

✓ FCC: The Federal Communications Commission. The agency which enforces the rules for the amateur radio service in the United States.

✓ Part 97: The part of the FCC rules which contain the rules and regulations which govern the amateur radio service.

✓ Question: Which of the following stations transmits signals over the air from a remote receive site to a repeater for retransmission?
Answer: Auxiliary Station

Notes:

Lesson 2
The T1B Section
Authorized Frequencies

T1B01 [97.3(a)(28)] What is the ITU?
Answer: *A United Nations agency for information and communication technology issues.*

ITU stands for International Telecommunications Union. The ITU is a United Nations organization that coordinates the world's telecommunication activities. This helps prevent international incidents where one country may be interfering with the other's communications activities. One of the ways the ITU helps to maintain the good order of international communication is it divides the globe into regions, which we will get to in the next question.

T1B02 North American amateur stations are located in which ITU region?
Answer: *Region 2.*

This is a good one to memorize. The International Telecommunications Union divides the world into four regions. We just happen to fall into region 2.

T1B03 [97.301(a)] Which frequency is within the 6 meter band?
Answer: *52.525 MHz.*

Technician, General, and Extra class licensees all have the same privileges on the 6 meter band. The 6 meter band runs from 50 MHz to 54 MHz. The answer, 52.525 MHz, falls between 50 MHz and 54 MHz and therefore is within the 6 meter band. I listed the bands and frequencies you will need to memorize for the exam on the Study Sheet at the end of the lesson.

Before we get too far into frequencies, it is important to talk a little about the relationship between bands and frequencies. Radio signals travel in waves. Each radio wave has a specific length, or wavelength, which determines its frequency. For our purposes, these wavelengths are measured using the metric system (meters, centimeters, etc.). All radio waves travel at a set speed, the speed of light. The frequency of these radio waves is determined by how many radio waves pass a point in one second. The longer the wavelength, the less radio waves pass a point during a second, and the lower the frequency. As frequency increases, wavelength decreases. As wavelength increases, frequency decreases. Amateurs will often categorize groups, or bands, of frequencies using their wavelength. To use the above answer as an example, 52.525 MHz has a wavelength of roughly 6 meters and is categorized as being part of the 6 meter band. There will be more on this later.

T1B04 [97.301(a)] Which amateur band are you using when your station is transmitting on 146.52 MHz?
Answer: *2 meter band.*

Same idea as the previous question. Amateur privileges on the 2 meter band extend from 144 MHz to 148 MHz. 146.52 MHz falls within 144 MHz and 148 MHz thus making it part of the 2 meter band. Another one you will need to memorize.

T1B05 [97.301(a)] Which 70 cm frequency is authorized to a Technician Class license holder operating in ITU Region 2?
Answer: *443.350 MHz.*

If you remember from earlier, North American amateur stations fall within ITU region 2. This is us! The next thing you need to know to answer this question is what your 70 cm privileges will be as a Technician Class license holder. They are 420 MHz to 450 MHz. Not only does 443.350 MHz fall within the 70 cm amateur privileges, it is the only answer that even comes close!

T1B06 [97.301(a)] Which 23 cm frequency is authorized to a Technician Class operator license?
Answer: *1296 MHz.*

The 23 cm amateur frequencies for Technician, General, and Extra class operators are 1240 MHz to 1300 MHz. Another one you just need to memorize.

T1B07 [97.301(a)] What amateur band are you using if you are transmitting on 223.50 MHz?
Answer: *1.25 meter band.*

The 1.25 meter amateur frequencies are 219 MHz to 220 MHz and 222 MHz to 225 MHz. As you can see, the band is split into two groups. 219 MHz to 220 MHz is solely for amateur fixed digital message forwarding systems. You can you use other modes between 222 MHz and 225 MHz.

T1B08 [97.303] What do the FCC rules mean when an amateur frequency band is said to be available on a secondary basis?
Answer: *Amateurs may not cause harmful interference to primary users.*

If you are allowed to use a band on a secondary basis it means you are not allowed to interfere with stations who have primary use on that band. You can think of it as being a guest in someone else's house. You may be able to watch TV and sit on the couch, but you will have to leave when the owners want to use them. You are essentially sharing the frequencies with the primary stations. However, if the primary station wants to use the frequency, they have the right of way. You have to stay out of their way and not interfere.

T1B09 [97.101(a)] Why should you not set your transmit frequency to be exactly at the edge of an amateur band or sub-band?
Answer: *To allow for calibration error in the transmitter frequency display, so that modulation sidebands do not extend beyond the band edge, and to allow for transmitter frequency drift.*

This is an "all of the above" question on the exam. The point this question is trying to make is that your entire transmitted signal must stay completely within the authorized frequency limits allocated to your license class. You never want to transmit on a frequency on the very edge of an authorized amateur band. Several things can happen. Your transmitter frequency display may be off by a few Hz which might cause all or part of your signal to go outside the range. Your signal may produce unwanted (and sometimes unnoticed) modulation or harmonics that can end up outside the band. Some transmitters may drift a bit as well. This is when your transmitter mysteriously starts to move up or down in frequency without your permission. This happens a lot in older transmitter models which may need some time for the circuits to warm up before it decides to stick to the frequency you want. You need to make sure your entire signal stays within the amateur band!

T1B10 [97.305(c)] Which of the bands available to Technician Class operators have mode-restricted sub-bands?
Answer: *The 6 meter, 2 meter, and 1.25 meter bands.*

A mode restricted sub-band is a portion of a band that you are only allowed to transmit with a certain mode. There are only three bands available to Technician Class operators that fall into this category: 6 meter, 2 meter, and 1.25 meter. 6 meter and 2 meter have a small portion where an operator can only transmit in CW mode, otherwise known as Morse Code. For 6 meters, this is from 50.0 MHz to 50.1 MHz. For 2 meters, it is from 144.0 MHz to 144.1 MHz. The 1.25 meter band has 219 MHz to 220 MHz restricted for fixed digital message forwarding systems. You are permitted to use whatever mode you would like in the rest of those bands.

T1B11 [97.305 (a)(c)] What emission modes are permitted in the mode-restricted sub-bands at 50.0 to 50.1 MHz and 144.0 to 144.1 MHz?
Answer: *CW only.*

We went over this in the previous question. CW stands for Continuous Wave and is essentially Morse Code. You are only allowed to use CW (Morse Code) in those areas of the 6 meter and 2 meter bands.

Quiz #2

(Find the answers on page 256)

1. _____ T1B01 [97.3(a)(28)] What is the ITU?
A. An agency of the United States Department of Telecommunications Management
B. A United Nations agency for information and communication technology issues
C. An independent frequency coordination agency
D. A department of the FCC

2. _____ T1B02 North American amateur stations are located in which ITU region?
A. Region 1
B. Region 2
C. Region 3
D. Region 4

3. _____ T1B03 [97.301(a)] Which frequency is within the 6 meter band?
A. 49.00 MHz
B. 52.525 MHz
C. 28.50 MHz
D. 222.15 MHz

4. _____ T1B04 [97.301(a)] Which amateur band are you using when your station is transmitting on 146.52 MHz?
A. 2 meter band
B. 20 meter band
C. 14 meter band
D. 6 meter band

5. _____ T1B05 [97.301(a)] Which 70 cm frequency is authorized to a Technician Class license holder operating in ITU Region 2?
A. 53.350 MHz
B. 146.520 MHz
C. 443.350 MHz
D. 222.520 MHz

6. _____ T1B06 [97.301(a)] Which 23 cm frequency is authorized to a Technician Class operator license?
A. 2315 MHz
B. 1296 MHz
C. 3390 MHz
D. 146.52 MHz

7. _____ T1B07 [97.301(a)] What amateur band are you using if you are transmitting on 223.50 MHz?
A. 15 meter band
B. 10 meter band
C. 2 meter band
D. 1.25 meter band

8. _____ T1B08 [97.303] What do the FCC rules mean when an amateur frequency band is said to be available on a secondary basis?
A. Secondary users of a frequency have equal rights to operate
B. Amateurs are only allowed to use the frequency at night
C. Amateurs may not cause harmful interference to primary users
D. Secondary users are not allowed on amateur bands

9. _____ T1B09 [97.101(a)] Why should you not set your transmit frequency to be exactly at the edge of an amateur band or sub-band?
A. To allow for calibration error in the transmitter frequency display
B. So that modulation sidebands do not extend beyond the band edge
C. To allow for transmitter frequency drift
D. All of these choices are correct

10. _____ T1B10 [97.305(c)] Which of the bands available to Technician Class operators have mode-restricted sub-bands?
A. The 6 meter, 2 meter, and 70 cm bands
B. The 2 meter and 13 cm bands
C. The 6 meter, 2 meter, and 1.25 meter bands
D. The 2 meter and 70 cm bands

11. _____ T1B11 [97.305 (a)(c)] What emission modes are permitted in the mode-restricted sub-bands at 50.0 to 50.1 MHz and 144.0 to 144.1 MHz?
A. CW only
B. CW and RTTY
C. SSB only
D. CW and SSB

Lesson 2 Study Sheet

✓ There are more frequency privileges allocated for Technician Class operators than the ones listed below. The below bands and frequencies are the just the ones you will need to study for the exam. For a complete listing of Technician Class privileges see Figure 1 on page 3.

6 meters

50-54MHz

RTTY, data, phone, and image from 50.1 MHz to 54 MHz. CW only from 50 MHz to 50.1 MHz.

2 meters

144-148 MHz

RTTY, data, phone, and image from 144.1 MHz to 148 MHz. CW only from 144 MHz to 144.1 MHz.

1.25 meters

219 MHz to 220 MHz and 222 MHz to 225 MHz

219 MHz to 220 MHz is fixed digital message forwarding system only.

RTTY, data, phone, and image from 222 MHz to 225 MHz.

70 cm

420 MHz to 450 MHz RTTY, data, phone, and image.

23 cm

1240 MHz to 1300 MHz all modes.

✓ ITU: A United Nations agency for information and communication technology issues.

✓ North America is located in ITU region 2.

Notes:

Lesson 3
The T1C Section
Station Call Signs and Licenses

T1C01 [97.3(a)(11)(iii)] Which type of call sign has a single letter in both the prefix and suffix?
Answer: *Special event.*

The FCC issues out several different types of call signs. A special event call sign is assigned by the FCC to identify stations participating in a special event. These special event call signs consist of a single letter preface , followed with a number, followed by a single letter suffix. Basically, a number sandwiched between two letters. K1A is an example of a special event call sign.

T1C02 Which of the following is a valid US amateur radio station call sign?
Answer: *W3ABC*

Most US amateur call signs will begin with a 'K' or 'W'. This is not completely the case, but making this assumption will prevent confusion on the exam. This will be followed within the next two digits by a single number. Following this number will be a series of two or three letters. You can tell a lot from a call sign. The first letter will tell you that it is a US amateur. The number will tell you where in the United States the amateur's license is registered. For instance, a '4' tells you that the ham is probably located in the Mid-Atlantic states. A '1' will tell you the station is probably in the New England area. The remainder of the letters just further help identify the amateur.

T1C03 [97.117] What types of international communications are permitted by an FCC-licensed amateur station?
Answer: *Communications incidental to the purposes of the amateur service and remarks of a personal character.*

If you think back to the first lesson, amateur radio is meant for those who have a personal interest in radio technology without an interest in using their amateur privileges to make money. Especially when engaged in international communications, you want to keep the conversation to the spirit of amateur radio. You should think of yourself as an ambassador for US amateurs!

T1C04 When are you allowed to operate your amateur station in a foreign country?
Answer: *When the foreign country authorizes it.*

You are allowed to operate in a foreign country as long as that country allows it. Getting their permission is another story. Some countries have no problem allowing US amateurs to operate within their country, others require you be granted a form of temporary license, and others flat out don't let you. The ARRL's website (www.arrl.org) has information about operating in some of the more frequently traveled countries. The US State Department can help as well. Whatever you do, make sure you are in compliance with that country's laws before you hit transmit. Some countries have strict penalties for breaking their laws!

T1C05 [97.303(h)] What must you do if you are operating on the 23 cm band and learn that you are interfering with a radiolocation station outside the United States?
Answer: *Stop operating or take steps to eliminate the harmful interference.*

This is just polite and there are very few times when you find yourself interfering with another station that you shouldn't stop and take steps to eliminate the problem. However, this situation is specifically addressed by the Part 97 rules. There is a possibility that you may hear radiolocation stations transmitting outside the United States on the 23 cm band. Do not transmit near the frequency you hear them on and stay out of their way!

T1C06 [97.5(a)(2)] From which of the following may an FCC-licensed amateur station transmit, in addition to places where the FCC regulates communications?
Answer: *From any vessel or craft located in international waters and documented or registered in the United States.*

Your FCC privileges are not limited to just within the United States. If you are on a boat in international waters and that boat is registered in the United States, then you can transmit!

T1C07 [97.23] What may result when correspondence from the FCC is returned as undeliverable because the grantee failed to provide the correct mailing address?
Answer: *Revocation of the station license or suspension of the operator license.*

Make sure your personal information is up to date in the FCC records. If you move to a new address and forget to update your records with the FCC, you may be in for a surprise. If the FCC needs to get in touch with you and can't because your address is wrong, they will suspend or revoke your license!

T1C08 [97.25] What is the normal term for an FCC-issued primary station/operator license grant?
Answer: *Ten years.*

An amateur license is good for a decade. This one you will need to memorize.

T1C09 [97.21(a)(b)] What is the grace period following the expiration of an amateur license within which the license may be renewed?
Answer: *Two years.*

Your license is good for ten years. If for some reason you forget to renew, you have an additional two years to renew your license before you have to retake the exam and start again from scratch. By the way, you are not allowed to operate during that two year grace period. Just because you are still within the grace period does not mean you have a valid license!

T1C10 97.5a] How soon may you operate a transmitter on an amateur service frequency after you pass the examination required for your first amateur radio license?
Answer: *As soon as your name and call sign appear in the FCC's ULS database.*

ULS stands for Universal Licensing System and is the system used by the FCC to manage all the various licenses they deal with. Gone are the days when you had to wait for your license in the mail before you could start operating. It is online and fairly user friendly. As soon as your call sign appears in the database, it means the FCC has processed everything they need to and you can start hitting the transmit button!

T1C11 [97.21(b)] If your license has expired and is still within the allowable grace period, may you continue to operate a transmitter on amateur service frequencies?
Answer: *No, transmitting is not allowed until the ULS database shows that the license has been renewed.*

Just because you are still within the grace period for renewal does not mean you are authorized to transmit. Just like getting your first license, you don't have to wait on the mail. You just have to wait for the ULS database to say your license is renewed.

Quiz #3

(Find the answers on page 256)

1. _____ T1C01 [97.3(a)(11)(iii)] Which type of call sign has a single letter in both the prefix and suffix?
A. Vanity
B. Sequential
C. Special event
D. In-memoriam

2. _____ T1C02 Which of the following is a valid US amateur radio station call sign?
A. KMA3505
B. W3ABC
C. KDKA
D. 11Q1176

3. _____ T1C03 [97.117] What types of international communications are permitted by an FCC-licensed amateur station?
A. Communications incidental to the purposes of the amateur service and remarks of a personal character
B. Communications incidental to conducting business or remarks of a personal nature
C. Only communications incidental to contest exchanges, all other communications are prohibited
D. Any communications that would be permitted on an international broadcast station

4. _____ T1C04 When are you allowed to operate your amateur station in a foreign country?
A. When the foreign country authorizes it
B. When there is a mutual agreement allowing third party communications
C. When authorization permits amateur communications in a foreign language
D. When you are communicating with non-licensed individuals in another country

5. _____ T1C05 [97.303(h)] What must you do if you are operating on the 23 cm band and learn that you are interfering with a radiolocation station outside the United States?
A. Stop operating or take steps to eliminate the harmful interference
B. Nothing, because this band is allocated exclusively to the amateur service
C. Establish contact with the radiolocation station and ask them to change frequency
D. Change to CW mode, because this would not likely cause interference

6. _____ T1C06 [97.5(a)(2)] From which of the following may an FCC-licensed amateur station transmit, in addition to places where the FCC regulates communications?
A. From within any country that belongs to the International Telecommunications Union
B. From within any country that is a member of the United Nations
C. From anywhere within in ITU Regions 2 and 3
D. From any vessel or craft located in international waters and documented or registered in the United States

7. _____ T1C07 [97.23] What may result when correspondence from the FCC is returned as undeliverable because the grantee failed to provide the correct mailing address?
A. Fine or imprisonment
B. Revocation of the station license or suspension of the operator license
C. Require the licensee to be re-examined
D. A reduction of one rank in operator class

8. _____ T1C08 [97.25] What is the normal term for an FCC-issued primary station/operator license grant?
A. Five years
B. Life
C. Ten years
D. Twenty years

9. _____ T1C09 [97.21(a)(b)] What is the grace period following the expiration of an amateur license within which the license may be renewed?
A. Two years
B. Three years
C. Five years
D. Ten years

10. _____ T1C10 [97.5a] How soon may you operate a transmitter on an amateur service frequency after you pass the examination required for your first amateur radio license?
A. Immediately
B. 30 days after the test date
C. As soon as your name and call sign appear in the FCC's ULS database
D. You must wait until you receive your license in the mail from the FCC

11. _____ T1C11 [97.21(b)] If your license has expired and is still within the allowable grace period, may you continue to operate a transmitter on amateur service frequencies?
A. No, transmitting is not allowed until the ULS database shows that the license has been renewed
B. Yes, but only if you identify using the suffix "GP"
C. Yes, but only during authorized nets
D. Yes, for up to two years

Lesson 3 Study Sheet

✓ A special event call sign is a single letter prefix, followed with a number, followed by a single letter suffix. For instance, K1A.

✓ An example of a US amateur call sign is K3ABC.

✓ An FCC issued amateur radio license is good for 10 years.

✓ There is a two year grace period to renew an amateur license after the license has expired. The amateur is not allowed to operate until the license is listed as renewed in the FCC's ULS database.

Notes:

Lesson 4
The T1D Section
Authorized and Prohibited Transmissions

T1D01 [97.111(a)(1)] With which countries are FCC-licensed amateur stations prohibited from exchanging communications?
Answer: *Any country whose administration has notified the ITU that it objects to such communications.*

You can talk to most foreign amateur stations. There are exceptions. If a country has informed the International Telecommunications Union (ITU) that they do not want their radio operators in their country talking to foreign amateurs, then they are off limits. North Korea would be a good example of this. There is a trick answer on the exam! Remember, the country must notify the ITU. Simply notifying the United Nations is not enough. You are looking for the best answer!

T1D02 [97.111(a)(5)] On which of the following occasions may an FCC-licensed amateur station exchange messages with a U.S. military station?
Answer: *During an Armed Forces Day Communications Test.*

Of the possible answers, the only time you are allowed to talk to a U.S. Military station is during an Armed Forces Day communications test. This answer makes the most sense. It has the words "Military" and "Communications" in it. Keep it simple!

T1D03 [97.113(a)(4), 97.211(b), 97.217] When is the transmission of codes or ciphers allowed to hide the meaning of a message transmitted by an amateur station?
Answer: *Only when transmitting control commands to space stations or radio control craft.*

If you think about it, this makes sense. There are two amateur activities that could cause a lot of damage if tampered with by unauthorized transmissions: the control of satellites and control of radio controlled craft. Imagine what would happen if an amateur accidentally sends a signal that shuts down an amateur satellite. Even worse, you are flying you radio controlled airplane and another ham inadvertently sends a signal that nose dives your plane into a playground! This is the reason that these hams who work with amateur satellites or radio control aircraft are allowed to send coded signals! However, this is the only exception!

T1D04 [97.113(a)(4), 97.113(e)] What is the only time an amateur station is authorized to transmit music?
Answer: *When incidental to an authorized retransmission of manned spacecraft communications.*

The only time I can think this would be applicable is if you are talking to a ham astronaut on the Space Shuttle or International Space Station and they play the crew's musical morning wake-up call. One way to help remember this answer is to think of Frank Sinatra singing *Fly Me To the Moon*!

T1D05 [97.113(a)(3)] When may amateur radio operators use their stations to notify other amateurs of the availability of equipment for sale or trade?
Answer: *When the equipment is normally used in an amateur station and such activity is not conducted on a regular basis.*

Ham radio is not a marketplace! Remember, amateurs are not in it for monetary gain. However, if you have a piece of amateur radio equipment that you want to sell, you can coordinate selling it on the air. The point is to keep the selling of equipment infrequent, so do not make a habit of it!

T1D06 [97.113(a)(4)] Which of the following types of transmissions are prohibited?
Answer: *Transmissions that contain obscene or indecent words or language.*

This is the only possible answer that makes sense as well as the only one with a negative tone. Obscene or indecent language is absolutely forbidden! Amateurs are courteous to one another. Also, there are many minors who are amateurs as well. Keep the conversation clean and G rated!

T1D07 [97.113(f)] When is an amateur station authorized to automatically retransmit the radio signals of other amateur stations?
Answer: *When the signals are from an auxiliary, repeater, or space station.*

Auxiliary, repeater, and most space stations automatically retransmit signals. For the most part, these stations allow an amateur operator to greatly increase the range of their signal. For instance, a repeater can take a relatively low powered signal from your handheld radio and retransmit it at a much higher power. A little information may help rule out some of the other possible answers. A few of the answers give Earth stations and beacons as possible answers. An Earth station can be any station on Earth which is much too broad and not all Earth stations are authorized to retransmit signals. A beacon station only sends information, it does not retransmit anything.

T1D08 [97.113] When may the control operator of an amateur station receive compensation for operating the station?
Answer: *When the communication is incidental to classroom instruction at an educational institution.*

Amateur radio teachers can be compensated for running a station as part of classroom instruction. Also, club employees may be compensated for performing their duties as well. For the exam, a teacher who is teaching students about amateur radio may receive payment for operating an amateur station.

T1D09 [97.113(b)] Under which of the following circumstances are amateur stations authorized to transmit signals related to broadcasting, program production, or news gathering, assuming no other means is available?
Answer: *Only where such communications directly relate to the immediate safety of human life or protection of property.*

If no other means is available, you may retransmit news or commercial broadcasts if human life or property is at risk. If someone is about to lose their life or property, you can send them whatever information necessary to help them. Be sure to remember this is allowed if no other means are available! This means amateur radio is all that is left to assist those in danger! Otherwise, retransmitting broadcasts are unauthorized.

T1D10 [97.3(a)(10)] What is the meaning of the term broadcasting in the FCC rules for the amateur services?
Answer: *Transmissions intended for reception by the general public.*

Broadcasting is intended for everyone. Think big picture! It is a general term for all types of radio services, not only ham radio. Primarily, this definition is directed towards TV and radio stations, otherwise known as broadcast stations, which serve the general public.

T1D11 [97.113(a)(5)] Which of the following types of communications are permitted in the Amateur Radio Service?
Answer: *Brief transmissions to make station adjustments.*

You are allowed to make brief transmissions to adjust your equipment. An example of this would be a short transmission to check to see if your transmitter is tuned to your antenna. Two of the other possible answers deal with retransmission of entertainment material which you already know is not authorized by the FCC. The other incorrect answer deals with using amateur radio for communications that could reasonably be furnished by another radio service. This would be something similar to using amateur radio for police communications when their is a police band available. These types of communications are also not authorized.

Quiz #4

(Find the answers on page 256)

1. _____ T1D01 [97.111(a)(1)] With which countries are FCC-licensed amateur stations prohibited from exchanging communications?
A. Any country whose administration has notified the ITU that it objects to such communications
B. Any country whose administration has notified the United Nations that it objects to such communications
C. Any country engaged in hostilities with another country
D. Any country in violation of the War Powers Act of 1934

2. _____ T1D02 [97.111(a)(5)] On which of the following occasions may an FCC-licensed amateur station exchange messages with a U.S. military station?
A. During an Armed Forces Day Communications Test
B. During a Memorial Day Celebration
C. During an Independence Day celebration
D. During a propagation test

3. _____ T1D03 [97.113(a)(4), 97.211(b), 97.217] When is the transmission of codes or ciphers allowed to hide the meaning of a message transmitted by an amateur station?
A. Only during contests
B. Only when operating mobile
C. Only when transmitting control commands to space stations or radio control craft
D. Only when frequencies above 1280 MHz are used

4. _____ T1D04 [97.113(a)(4), 97.113(e)] What is the only time an amateur station is authorized to transmit music?
A. When incidental to an authorized retransmission of manned spacecraft communications
B. When the music produces no spurious emissions
C. When the purpose is to interfere with an illegal transmission
D. When the music is transmitted above 1280 MHz

5. _____ T1D05 [97.113(a)(3)] When may amateur radio operators use their stations to notify other amateurs of the availability of equipment for sale or trade?
A. When the equipment is normally used in an amateur station and such activity is not conducted on a regular basis
B. When the asking price is $100.00 or less
C. When the asking price is less than its appraised value
D. When the equipment is not the personal property of either the station licensee or the control operator or their close relatives

6. _____ T1D06 [97.113(a)(4)] Which of the following types of transmissions are prohibited?
A. Transmissions that contain obscene or indecent words or language
B. Transmissions to establish one-way communications
C. Transmissions to establish model aircraft control
D. Transmissions for third party communications

7. _____ T1D07 [97.113(f)] When is an amateur station authorized to automatically retransmit the radio signals of other amateur stations?
A. When the signals are from an auxiliary, beacon, or Earth station
B. When the signals are from an auxiliary, repeater, or space station
C. When the signals are from a beacon, repeater, or space station
D. When the signals are from an Earth, repeater, or space station

8. _____ T1D08 [97.113]When may the control operator of an amateur station receive compensation for operating the station?
A. When engaging in communications on behalf of their employer
B. When the communication is incidental to classroom instruction at an educational institution
C. When re-broadcasting weather alerts during a RACES net
D. When notifying other amateur operators of the availability for sale or trade of apparatus

9. _____ T1D09 [97.113(b)] Under which of the following circumstances are amateur stations authorized to transmit signals related to broadcasting, program production, or news gathering, assuming no other means is available?
A. Only where such communications directly relate to the immediate safety of human life or protection of property
B. Only when broadcasting communications to or from the space shuttle.
C. Only where noncommercial programming is gathered and supplied exclusively to the National Public Radio network
D. Only when using amateur repeaters linked to the Internet

10. _____ T1D10 [97.3(a)(10)] What is the meaning of the term broadcasting in the FCC rules for the amateur services?
A. Two-way transmissions by amateur stations
B. Transmission of music
C. Transmission of messages directed only to amateur operators
D. Transmissions intended for reception by the general public

11. _____ T1D11 [97.113(a)(5)] Which of the following types of communications are permitted in the Amateur Radio Service?
A. Brief transmissions to make station adjustments
B. Retransmission of entertainment programming from a commercial radio or TV station
C. Retransmission of entertainment material from a public radio or TV station
D. Communications on a regular basis that could reasonably be furnished alternatively through other radio services

Lesson 4 Study Sheet

✓ You may talk to amateurs in any foreign country unless that country has notified the ITU that those types of communications are prohibited.

✓ An amateur operator may communicate with a U.S. Military station only during an Armed Forces Day Communications Test.

✓ You may only transmit codes and ciphers when controlling space stations and radio control craft.

✓ Broadcasting is sending transmissions intended for the general public.

Notes:

Lesson 5
The T1E Section
Control Operators and Control Types

T1E01 [97.7(a)] When must an amateur station have a control operator?
Answer: *Only when the station is transmitting.*

A control operator is an amateur operator who is responsible for the transmissions from a station. The control operator is not necessarily the licensee of that station, but may be appointed by the licensee to operate the station and ensure that the station's transmissions are in compliance with the FCC. The control operator is the individual who is controlling the transmissions of the station. There does not need to be a control operator if the station is not transmitting.

T1E02 [97.7(a)] Who is eligible to be the control operator of an amateur station?
Answer: *Only a person for whom an amateur operator/primary station license grant appears in the FCC database or who is authorized for alien reciprocal operation.*

Within the United States, a control operator must be a licensed amateur operator. If the control operator happens to be a guest from a foreign country, they must be authorized by the FCC to transmit on the amateur frequencies within the U.S. In other words, the foreign amateur must be authorized for alien reciprocal operation. Age and citizenship does not matter.

T1E03 [97.103(b)] Who must designate the station control operator?
Answer: *The station licensee.*

The station licensee is the one the FCC will ultimately hold responsible for the proper operation of an amateur station. The station licensee may allow another amateur to be the control operator of the licensee's station. This must be documented. You should document in your logbook the who, when, and how long someone other than yourself is the control operator of your station. Unless it is documented, the FCC assumes you are the control operator of your station! If you have a guest amateur over who is sends a prohibited transmission, you don't want to be the only one holding the bag when the Feds come!

T1E04 [97.103(b)] What determines the transmitting privileges of an amateur station?
Answer: *The class of operator license held by the control operator.*

Regardless who the licensee of the station is, the control operator's privileges determine what transmitting privileges the amateur station can operate.

T1E05 [97.3(a)(14)] What is an amateur station control point?
Answer: *The location at which the control operator function is performed.*

This is wherever the control operator has control of the station transmitting. Think of this as the place where you can hit the "panic" button and stop transmitting if things get out of hand. For the most part, this is in front of the radio. However, there are certain types of stations where the control operator may not be right in front of the radio but can remotely control the station or not be at the control point at all. More on this in the next question.

T1E06 [97.109(d)] Under which of the following types of control is it permissible for the control operator to be at a location other than the control point?
Answer: *Automatic control.*

There are three types of control permitted by the FCC. They are local control, remote control, and automatic control. For local control, the control operator and control point are at the transmitter. This is the ham sitting right in front of the radio. Any station can be remotely controlled. You are allowed to control a transmitter remotely. Some radios can be controlled remotely (remote control) by computers through the internet or handheld radios which may not be at the same location as the transmitter. For remote control, the control operator and control point is that handheld radio or computer which is controlling the transmitter. Any station may be remotely controlled. Automatic control is a station which does not need a control operator at the control point while the station is transmitting. The station transmits automatically per specific parameters. Repeaters are examples of stations which operate under automatic control. Only certain types of stations, such as repeaters and beacons, are authorized to operate under automatic control.

T1E07 [97.103(a)] When the control operator is not the station licensee, who is responsible for the proper operation of the station?
Answer: *The control operator and the station licensee are equally responsible.*

You will share the blame! The control operator is responsible for ensuring that the station is transmitting in compliance with FCC Part 97 rules and the station licensee is responsible for making sure the control operator knows what he or she is doing. If the control operator breaks the rules, both the control operator and licensee are held accountable by the FCC.

T1E08 [97.3(a)] What type of control is being used for a repeater when the control operator is not present at a control point?
Answer: *Automatic control.*

Sound familiar. If a station is legally operating when the control operator is not present at the control point, then the station is under automatic control.

T1E09 [97.109(a)] What type of control is being used when transmitting using a handheld radio?
Answer: *Local control.*

When you hit the push-to-talk button on your handheld radio, you are the control operator of that station. Since you are present at the transmitter, that handheld is under local control.

T1E10 [97.3] What type of control is used when the control operator is not at the station location but can indirectly manipulate the operating adjustments of a station?
Answer: *Remote.*

If you are controlling a transmitter located at a different location via a computer or some other device, this is remote control.

T1E11 [97.103(a)] Who does the FCC presume to be the control operator of an amateur station, unless documentation to the contrary is in the station records?
Answer: *The station licensee.*

It is important to document if you let someone else act as the control operator of your station. Otherwise, the FCC will assume you are the control operator.

Quiz #5

(Find the answers on page 256)

1. _____ T1E01 [97.7(a)] When must an amateur station have a control operator?
A. Only when the station is transmitting
B. Only when the station is being locally controlled
C. Only when the station is being remotely controlled
D. Only when the station is being automatically controlled

2. _____ T1E02 [97.7(a)] Who is eligible to be the control operator of an amateur station?
A. Only a person holding an amateur service license from any country that belongs to the United Nations
B. Only a citizen of the United States
C. Only a person over the age of 18
D. Only a person for whom an amateur operator/primary station license grant appears in the FCC database or who is authorized for alien reciprocal operation

3. _____ T1E03 [97.103(b)] Who must designate the station control operator?
A. The station licensee
B. The FCC
C. The frequency coordinator
D. The ITU

4. _____ T1E04 [97.103(b)] What determines the transmitting privileges of an amateur station?
A. The frequency authorized by the frequency coordinator
B. The class of operator license held by the station licensee
C. The highest class of operator license held by anyone on the premises
D. The class of operator license held by the control operator

5. _____ T1E05 [97.3(a)(14)] What is an amateur station control point?
A. The location of the station's transmitting antenna
B. The location of the station transmitting apparatus
C. The location at which the control operator function is performed
D. The mailing address of the station licensee

6. _____ T1E06 [97.109(d)] Under which of the following types of control is it permissible for the control operator to be at a location other than the control point?
A. Local control
B. Automatic control
C. Remote control
D. Indirect control

7. _____ T1E07 [97.103(a)] When the control operator is not the station licensee, who is responsible for the proper operation of the station?
A. All licensed amateurs who are present at the operation
B. Only the station licensee
C. Only the control operator
D. The control operator and the station licensee are equally responsible

8. _____ T1E08 [97.3(a)] What type of control is being used for a repeater when the control operator is not present at a control point?
A. Local control
B. Remote control
C. Automatic control
D. Unattended

9. _____ T1E09 [97.109(a)] What type of control is being used when transmitting using a handheld radio?
A. Radio control
B. Unattended control
C. Automatic control
D. Local control

10. _____ T1E10 [97.3] What type of control is used when the control operator is not at the station location but can indirectly manipulate the operating adjustments of a station?
A. Local
B. Remote
C. Automatic
D. Unattended

11. _____ T1E11 [97.103(a)] Who does the FCC presume to be the control operator of an amateur station, unless documentation to the contrary is in the station records?
A. The station custodian
B. The third party participant
C. The person operating the station equipment
D. The station licensee

Lesson 5 Study Sheet

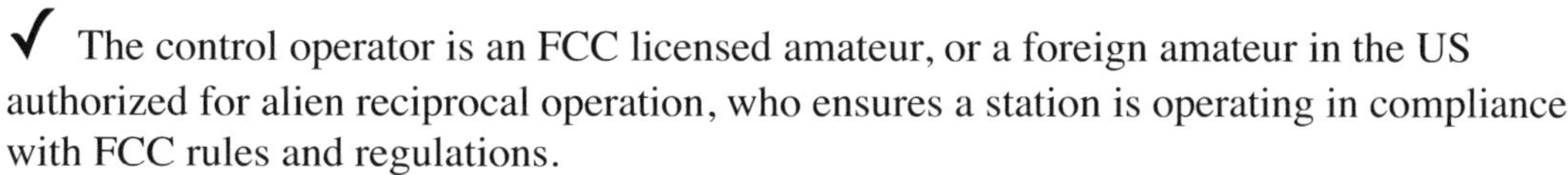

✓ The control operator is an FCC licensed amateur, or a foreign amateur in the US authorized for alien reciprocal operation, who ensures a station is operating in compliance with FCC rules and regulations.

✓ The control operator only needs to be present when the station is transmitting.

✓ The control point is the point where the control operator function is performed.

✓ Local control: the control operator is directly at the control point of the station.

✓ Remote control: The control operator is manipulating the controls of the station remotely. The station may be in another location.

✓ Automatic control: The control operator does not need to be at the control point while the station is transmitting. Only certain types of stations may be under automatic control.

Notes:

Lesson 6
The T1F Section
Station Identification and Operation Standards

T1F01 What type of identification is being used when identifying a station on the air as "Race Headquarters"?
Answer: *Tactical call.*

A tactical call is a way of identifying a station or operator by their role in an event. Amateurs perform many public services. Amateurs are often asked to provide communications support in marathons and other types of races. "Race Headquarters" in this question refers to the station that is at race headquarters...simple enough. A tactical call is not assigned by the FCC. It is specific to the communications plan for the situation or event. A tactical call is used when a station or operator needs to be identified by more than just their call sign and helps identify the role they are fulfilling. It would be a lot more difficult for everyone to remember the call sign of the amateur coordinating communications at a race headquarters.

T1F02 [97.119 (a)] When using tactical identifiers, how often must your station transmit the station's FCC-assigned call sign?
Answer: *Every ten minutes.*

Regardless of whether you are using a tactical call, you must identify yourself with your call sign every ten minutes! Just because you are working a field event or performing a public service doesn't mean the FCC no longer needs to know who you are. When you are operating a station, you must identify yourself every ten minutes.

T1F03 [97.119(a)] When is an amateur station required to transmit its assigned call sign?
Answer: *At least every 10 minutes during and at the end of a contact.*

This is important! You must identify yourself with your call sign at least every ten minutes during a contact and at the end of every contact.

T1F04 [97.119(b)] Which of the following is an acceptable language for use for station identification when operating in a phone sub-band?
Answer: *The English language.*

You are allowed to speak any language you would like during the contact. However, as long as you are an FCC licensed amateur, you must send your call sign in English. Remember, you must identify your station call every ten minutes during a contact and at the end of every contact, in English!

T1F05 [97.119(b)] What method of call sign identification is required for a station transmitting phone signals?
Answer: *Send the call sign using CW or phone emission.*

The two ways you may send your call using phone or voice modes is either speak it or send it using Morse Code. You will often hear repeater stations identify the repeater with CW. Do not get tricked on this answer! When operating using phone signals, you can send your call sign using phone or CW!

T1F06 [97.119(c)] Which of the following formats of a self-assigned indicator is acceptable when identifying using a phone transmission?
Answer: *KL7CC stroke W3, KL7CC slant W3, and KL7CC slash W3.*

Self-assigned indicators are exactly what they say they are, self-assigned indicators. They can be used in a similar way to tactical calls to help identify the stations function in an event, or they may be used to help identify the station's location. There are many possibilities for sending a self-assigned indicator. The proper way to identify a self-assigned indicator is to send a "/" between the call sign and the self-assigned indicator. This is easy enough for CW or data modes. You simply type "/" or send the Morse Code for the character. For phone, you must say "slash," or a synonym for the word "slash." "Slant" and "stroke" can be considered synonyms for "slash." This is an "all of the above" question on the exam.

T1F07 [97.119(c)] Which of the following restrictions apply when appending a self-assigned call sign indicator?
Answer: *It must not conflict with any other indicator specified by the FCC rules or with any call sign prefix assigned to another country.*

Self-assigned indicators are intended to provide clarity to communications, not to confuse them. When using self-assigned indicator it is important to ensure that the indicator you are using is not already acknowledged by the FCC or a foreign country to mean something else than what you intend. For instance, you would not want to use "SOS" as a self-assigned indicator. A little extreme, but you get the point.

T1F08 [97.119(e)] When may a Technician Class licensee be the control operator of a station operating in an exclusive Extra Class operator segment of the amateur bands?
Answer: *Never.*

There is never an occasion! You must stick to the band privileges assigned to your license!

T1F09 [97.3(a)(39)] What type of amateur station simultaneously retransmits the signal of another amateur station on a different channel or channels?
Answer: *Repeater station.*

Repeater stations are extremely popular especially on the 2 meter and 70 cm bands. What a repeater station does is receive your signal on one frequency and retransmits it on another frequency. The retransmitted signal is usually of a much higher power than the received signal. This allows amateurs using a low power handheld radio to get a much longer range for their transmission.

T1F10 [97.205(g)] Who is accountable should a repeater inadvertently retransmit communications that violate the FCC rules?
Answer: *The control operator of the originating station.*

If you send an unauthorized transmission through a repeater, you are the one accountable! The control operator of the repeater is not held accountable even though the repeater retransmitted the unauthorized transmission. There is no immunity for the originating station just because the signal was sent through a repeater.

T1F11 [97.115(a)] To which foreign stations do the FCC rules authorize the transmission of non-emergency third party communications?
Answer: *Any station whose government permits such communications.*

Third party communications is the passing of messages on amateur frequencies from one amateur to another amateur for delivery to a third party. It is completely legal for U.S. amateurs to pass third party traffic through other U.S. amateurs. It can get very precarious very quickly when one of the amateur stations is located in another country. Before passing third party traffic to a foreign station, make sure that country has a third party agreement with the United States. Not all countries allow their amateurs to pass or receive messages on behalf of a third party. Foreign third party communications can be tricky, so proceed with caution!

T1F12 [97.5(b)(2)] How many persons are required to be members of a club for a club station license to be issued by the FCC?
Answer: *At least 4.*

Four amateur operators is all it takes for the FCC to issue a club station license! All you need to do is find three of your closest ham friends and the four of you can get a club station license from the FCC!

T1F13 [97.103(c)] When must the station licensee make the station and its records available for FCC inspection?
Answer: *Any time upon request by an FCC representative.*

It's a simple answer. The FCC can request your station's records whenever the FCC wants to. The FCC is king and they do not have time to mess around. If a representative from the FCC asks you for information about your station or its records, be polite and be quick.

Quiz #6

(Find the answers on page 256)

1. _____ T1F01 What type of identification is being used when identifying a station on the air as "Race Headquarters"?
A. Tactical call
B. Self-assigned designator
C. SSID
D. Broadcast station

2. _____ T1F02 [97.119 (a)] When using tactical identifiers, how often must your station transmit the station's FCC-assigned call sign?
A. Never, the tactical call is sufficient
B. Once during every hour
C. Every ten minutes
D. At the end of every communication

3. _____ T1F03 [97.119(a)] When is an amateur station required to transmit its assigned call sign?
A. At the beginning of each contact, and every 10 minutes thereafter
B. At least once during each transmission
C. At least every 15 minutes during and at the end of a contact
D. At least every 10 minutes during and at the end of a contact

4. _____ T1F04 [97.119(b)] Which of the following is an acceptable language for use for station identification when operating in a phone sub-band?
A. Any language recognized by the United Nations
B. Any language recognized by the ITU
C. The English language
D. English, French, or Spanish

5. _____ T1F05 [97.119(b)] What method of call sign identification is required for a station transmitting phone signals?
A. Send the call sign followed by the indicator RPT
B. Send the call sign using CW or phone emission
C. Send the call sign followed by the indicator R
D. Send the call sign using only phone emission

6. _____ T1F06 [97.119(c)] Which of the following formats of a self-assigned indicator is acceptable when identifying using a phone transmission?
A. KL7CC stroke W3
B. KL7CC slant W3
C. KL7CC slash W3
D. All of these choices are correct

7. _____ T1F07 [97.119(c)] Which of the following restrictions apply when appending a self-assigned call sign indicator?
A. It must be more than three letters and less than five letters
B. It must be less than five letters
C. It must start with the letters AA through AL, K, N, or W and be not less than two characters or more than five characters in length
D. It must not conflict with any other indicator specified by the FCC rules or with any call sign prefix assigned to another country

8. _____ T1F08 [97.119(e)] When may a Technician Class licensee be the control operator of a station operating in an exclusive Extra Class operator segment of the amateur bands?
A. Never
B. On Armed Forces Day
C. As part of a multi-operator contest team
D. When using a club station whose trustee is an Extra Class operator licensee

9. _____ T1F09 [97.3(a)(39)] What type of amateur station simultaneously retransmits the signal of another amateur station on a different channel or channels?
A. Beacon station
B. Earth station
C. Repeater station
D. Message forwarding station

10. _____ T1F10 [97.205(g)] Who is accountable should a repeater inadvertently retransmit communications that violate the FCC rules?
A. The control operator of the originating station
B. The control operator of the repeater
C. The owner of the repeater
D. Both the originating station and the repeater owner

11. _____ T1F11 [97.115(a)] To which foreign stations do the FCC rules authorize the transmission of non-emergency third party communications?
A. Any station whose government permits such communications
B. Those in ITU Region 2 only
C. Those in ITU Regions 2 and 3 only
D. Those in ITU Region 3 only

12. _____ T1F12 [97.5(b)(2)] How many persons are required to be members of a club for a club station license to be issued by the FCC?
A. At least 5
B. At least 4
C. A trustee and 2 officers
D. At least 2

13. _____ T1F13 [97.103(c)] When must the station licensee make the station and its records available for FCC inspection?
A. Any time upon request by an official observer
B. Any time upon request by an FCC representative
C. 30 days prior to renewal of the station license
D. 10 days before the first transmission

Lesson 6 Study Sheet

✓ A tactical call is an identifier used to identify a station's role in a special event. For instance, "Race Headquarters" is a tactical call.

✓ Amateurs must provide their station identification every ten minutes during a contact and at the end of a contact.

✓ When using phone, an amateur operator may provide station identification by sending their call sign using phone or CW.

✓ A repeater station simultaneously retransmits the signal of another amateur station on a different channel or channels.

✓ Third party communication is a message from one amateur control operator to another amateur station control operator on behalf of a third party.

Notes:

Lesson 7
The T2A Section
Station Operation

T2A01 What is the most common repeater frequency offset in the 2 meter band?
Answer: *Plus or minus 600 kHz.*

As we talked about in Lesson 6, repeater stations receive your signal on one frequency and retransmit it on another. The difference between the repeater's receiving frequency and its transmitting frequency is called the repeater's offset. In the case of the 2 meter band, the most common offset between a repeater's receiving frequency and transmitting frequency is plus or minus 600 kHz. Don't forget the "plus or minus." This is one you will need to memorize.

T2A02 What is the national calling frequency for FM simplex operations in the 70 cm band?
Answer: *446.000 MHz.*

Simplex operation simply means that you are transmitting and receiving on the same frequency. A calling frequency is a specific frequency where you find another ham to communicate with. Once you have made solid contact on the calling frequency, most hams will agree to move the conversation to another frequency to leave the calling frequency open. These calling frequencies are part of a voluntary band plan that has been agreed to by the amateur community. The calling frequency for the 70 cm band is 446 MHz. Another one you will need to memorize.

T2A03 What is a common repeater frequency offset in the 70 cm band?
Answer: *Plus or minus 5 MHz.*

This is essentially the same question as the first question in this lesson except this one is asking about the 70 cm band. It is important to note at this point that different bands will have different offset frequencies for their respective repeaters. The frequency offset between a 70 cm repeater's transmit and receive frequencies is plus or minus 5 MHz. Memorize.

T2A04 What is an appropriate way to call another station on a repeater if you know the other station's call sign?
Answer: *Say the station's call sign then identify with your call sign.*

When calling another station on a repeater, you always want to lead with the call sign of the station you are looking for and then follow it with your own. You can even add a "this is" in between the two calls if that makes it easier for you. If I were trying to get in touch with K3ABC, the call would go, "K3ABC, KE4GKP," or "K3ABC this is KE4GKP." Either way the call sign of the station being called leads the call sign of the station calling.

T2A05 What should you transmit when responding to a call of CQ?
Answer: *The other station's call sign followed by your call sign.*

CQ means "calling any station." It is what you say or send when you are looking for another amateur to make contact with. Responding to a CQ is very similar to calling a station on a repeater. You always lead with the other station's call sign followed by your own.

T2A06 What must an amateur operator do when making on-air transmissions to test equipment or antennas?
Answer: *Properly identify the transmitting station.*

Every time you send a signal on the air you must identify yourself with your call sign. It does not matter if the purpose of the transmission is to test equipment or make a contact. Every time you transmit you must identify the transmitting station. It is also polite to ask if the frequency is clear before sending a test transmission.

T2A07 Which of the following is true when making a test transmission?
Answer: *Station identification is required at least every ten minutes during the test and at the end.*

The same rules apply for station identification when you are sending test transmissions as when you are making a normal contact. You must identify yourself every ten minutes and at the end of the test transmission.

T2A08 What is the meaning of the procedural signal "CQ"?
Answer: *Calling any station.*

CQ means "calling any station." When you are looking for another ham to make contact with you send a series of CQs followed by your call sign. A CQ call would go something like, "CQ CQ CQ CQ this is K3ABC, K3ABC, K3ABC calling CQ and standing by." If another amateur, let's say KE4GKP, decides to answer your call, that station would respond, "K3ABC KE4GKP," or "K3ABC this is KE4GKP." From there you can exchange locations, signal reports, and strike up a conversation.

T2A09 What brief statement is often used in place of "CQ" to indicate that you are listening on a repeater?
Answer: *Say your call sign.*

Looking for a contact on a repeater is a little different procedure than the normal CQ call. On a repeater, you are sharing the same frequency with everyone who is using the repeater and it can get crowded quickly. Sending a standard CQ will take up much time and precious space on the repeater. The moral of the story is you want to leave your call to request contact with other stations as short as possible. To let other stations listening to a repeater know that you are looking for a contact, all you need to do is send your call sign. That's it.

T2A10 What is a band plan, beyond the privileges established by the FCC?
Answer: *A voluntary guideline for using different modes or activities within an amateur band.*

Band plans are guidelines to help maintain a courteous and efficient use of the amateur bands. They are a voluntary convention adopted by the amateur community. Band plans are not directed by the FCC. If you do not stick to the band plan you will not be in trouble with the FCC however, you may make many hams angry. It is better to stick to the band plan than to be labeled a "lid" (bad operator).

T2A11 [97.313(a)] What are the FCC rules regarding power levels used in the amateur bands?
Answer: *An amateur must use the minimum transmitter power necessary to carry out the desired communication.*

This is an important rule. You are only allowed to use the minimum power necessary to maintain the contact. You do not need a 1500 watt transmission to talk to the ham down the street when a half of a watt will work fine. There are several reasons behind this rule. It keeps stations from unnecessarily overpowering the airwaves not to mention limiting unwanted RF exposure to you and your neighbors!

Quiz #7

(Find the answers on page 256)

1. _____ T2A01 What is the most common repeater frequency offset in the 2 meter band?
A. Plus 500 kHz
B. Plus or minus 600 kHz
C. Minus 500 kHz
D. Only plus 600 kHz

2. _____ T2A02 What is the national calling frequency for FM simplex operations in the 70 cm band?
A. 146.520 MHz
B. 145.000 MHz
C. 432.100 MHz
D. 446.000 MHz

3. _____ T2A03 What is a common repeater frequency offset in the 70 cm band?
A. Plus or minus 5 MHz
B. Plus or minus 600 kHz
C. Minus 600 kHz
D. Plus 600 kHz

4. _____ T2A04 What is an appropriate way to call another station on a repeater if you know the other station's call sign?
A. Say "break, break" then say the station's call sign
B. Say the station's call sign then identify with your call sign
C. Say "CQ" three times then the other station's call sign
D. Wait for the station to call "CQ" then answer it

5. _____ T2A05 What should you transmit when responding to a call of CQ?
A. CQ followed by the other station’s call sign
B. Your call sign followed by the other station’s call sign
C. The other station’s call sign followed by your call sign
D. A signal report followed by your call sign

6. _____ T2A06 What must an amateur operator do when making on-air transmissions to test equipment or antennas?
A. Properly identify the transmitting station
B. Make test transmissions only after 10:00 p.m. local time
C. Notify the FCC of the test transmission
D. State the purpose of the test during the test procedure

7. _____ T2A07 Which of the following is true when making a test transmission?
A. Station identification is not required if the transmission is less than 15 seconds
B. Station identification is not required if the transmission is less than 1 watt
C. Station identification is required only if your station can be heard
D. Station identification is required at least every ten minutes during the test and at the end

8. _____ T2A08 What is the meaning of the procedural signal "CQ"?
A. Call on the quarter hour
B. A new antenna is being tested (no station should answer)
C. Only the called station should transmit
D. Calling any station

9. _____ T2A09 What brief statement is often used in place of "CQ" to indicate that you are listening on a repeater?
A. Say "Hello test" followed by your call sign
B. Say your call sign
C. Say the repeater call sign followed by your call sign
D. Say the letters "QSY" followed by your call sign

10. _____ T2A10 What is a band plan, beyond the privileges established by the FCC?
A. A voluntary guideline for using different modes or activities within an amateur band
B. A mandated list of operating schedules
C. A list of scheduled net frequencies
D. A plan devised by a club to use a frequency band during a contest

11. _____ T2A11 [97.313(a)] What are the FCC rules regarding power levels used in the amateur bands?
A. Always use the maximum power allowed to ensure that you complete the contact
B. An amateur may use no more than 200 watts PEP to make an amateur contact
C. An amateur may use up to 1500 watts PEP on any amateur frequency
D. An amateur must use the minimum transmitter power necessary to carry out the desired communication

Lesson 7 Study Sheet

✓ The most common frequency offset for a 2 meter band repeater is plus or minus 600 kHz.

✓ The most common frequency offset for a 70 cm band repeater is plus or minus 5 MHz.

✓ The national calling frequency for the 70 cm band is 446 MHz.

✓ CQ is a procedural signal meaning "calling any station."

Notes:

Lesson 8
The T2B Section
VHF/UHF Operating Practices

T2B01 What is the term used to describe an amateur station that is transmitting and receiving on the same frequency?
Answer: *Simplex communication.*

Simplex is the simplest way to operate because you are transmitting and receiving on the same frequency.

T2B02 What is the term used to describe the use of a sub-audible tone transmitted with normal voice audio to open the squelch of a receiver?
Answer: *CTCSS.*

CTCSS stands for Continuous Tone-Codes Squelch System. What CTCSS does is send a sub-audible tone to a muted receiver. The receiver will turn off the mute (the squelch opens) when it hears the correct sub-audible CTCSS tone. Essentially, CTCSS blocks out all unwanted signals from being received unless the transmitter is programmed with the correct CTCSS tone for that specific receiver. CTCSS is often used on club repeaters where the control operator wishes to limit the use of the repeater to only members of that club. The repeater's control operator will give the correct CTCSS tone to specific individuals which is then programmed into their transmitters. The CTCSS tone will turn on the receiver and allow just the authorized individuals access the repeater.

T2B03 Which of the following describes the muting of receiver audio controlled solely by the presence or absence of an RF signal?
Answer: *Carrier squelch.*

The squelch circuit allows you to avoid listening to static when your receiver is on but there is nobody transmitting. What a squelch does is mute a receiver until the receiver picks up a signal of a certain strength. Once the receiver picks up a signal, the squelch is kicked off, and the receiver is no longer muted. Almost all receivers have a squelch which can be adjusted to turn off at various RF signal strengths. Be careful not to mix up carrier squelch and CTCSS on the exam. The key words in this question are, "solely by the presence or absence of an RF signal." A carrier squelch is only controlled by the presence or absence of the RF signal.

T2B04 Which of the following common problems might cause you to be able to hear but not access a repeater even when transmitting with the proper offset?
Answer: *The repeater receiver requires audio tone burst for access, the repeater receiver requires a CTCSS tone for access, and the repeater receiver may require a DCS tone sequence for access.*

This is an "all of the above" question. There are several ways a repeater can control who has access to it. We already talked about CTCSS. Some repeater receivers require an audio tone burst or DCS tone sequence to activate the receiver. Audio tones and DCS tone sequences work similar to CTCSS and they may be compared to a combination on a lock. If you don't have the right audio burst or DCS tone sequence, you can't open the receiver.

T2B05 What determines the amount of deviation of an FM signal?
Answer: *The amplitude of the modulating signal.*

To answer this question correctly you need to know that FM stands for Frequency Modulation. The radio in your car is probably an FM radio. The way information is carried on an FM signal is by manipulating the frequency of a carrier wave by adding a modulating signal. It is the amplitude of the modulating signal waves that causes the frequency of the carrier wave to vary. This varying of the frequency of the carrier wave is called deviation.

T2B06 What happens when the deviation of an FM transmitter is increased?
Answer: *Its signal occupies more bandwidth.*

This question builds on the previous question. Bandwidth is the amount of space a signal occupies in the radio spectrum. What you need to remember for the exam is as the deviation of an FM signal increases, the bandwidth of the FM signal increases as well.

T2B07 What should you do if you receive a report that your station's transmissions are causing splatter or interference on nearby frequencies?
Answer: *Check your transmitter for off-frequency operation or spurious emissions.*

If you get a report that you are causing interference or splatter, stop what you are doing and check your equipment! The usual cause is something on your transmitter is not adjusted correctly or there could possibly a loose connection. Always strive to put your best foot forward when you are transmitting!

T2B08 What is the proper course of action if your station's transmission unintentionally interferes with another station?
Answer: *Properly identify your transmission and move to a different frequency.*

This answer is the common sense answer on the exam. If you accidentally interfere with another station's transmission, identify yourself with your call, apologize, and move on to a different frequency.

T2B09 [97.119(b)(2)] Which of the following methods is encouraged by the FCC when identifying your station when using phone?
Answer: *Use of a phonetic alphabet.*

The phonetic alphabet is internationally recognized and is often used when reception is bad. The phonetic alphabet are words which replace the individual alphabet letters. Alpha, Bravo, Charlie, Delta, etc. Spelling your call sign in phonetic letters allows it to be better understood in weak signal communications. For instance, KE4GKP would be Kilo Echo Four Gulf Kilo Papa.

T2B10 What is the "Q" signal used to indicate that you are receiving interference from other stations?
Answer: *QRM.*

QRM is manmade interference to your signal, usually from other stations. "Q" signals are from back in the days when Morse Code was used by everybody. The purpose behind them was to create accepted abbreviations for frequent radio jargon to avoid spelling it all out in code. It is a lot easier to send the code for QRM than it is to type out "I AM EXPERIENCING INTERFERENCE FROM OTHER STATIONS." This one you will need to memorize...sorry.

T2B11 What is the "Q" signal used to indicate that you are changing frequency?
Answer: *QSY.*

If you send or say QSY it means you are changing frequency. Another one to memorize.

Quiz #8

(Find the answers on page 256)

1. _____ T2B01 What is the term used to describe an amateur station that is transmitting and receiving on the same frequency?
A. Full duplex communication
B. Diplex communication
C. Simplex communication
D. Half duplex communication

2. _____ T2B02 What is the term used to describe the use of a sub-audible tone transmitted with normal voice audio to open the squelch of a receiver?
A. Carrier squelch
B. Tone burst
C. DTMF
D. CTCSS

3. _____ T2B03 Which of the following describes the muting of receiver audio controlled solely by the presence or absence of an RF signal?
A. Tone squelch
B. Carrier squelch
C. CTCSS
D. Modulated carrier

4. _____ T2B04 Which of the following common problems might cause you to be able to hear but not access a repeater even when transmitting with the proper offset?
A. The repeater receiver requires audio tone burst for access
B. The repeater receiver requires a CTCSS tone for access
C. The repeater receiver may require a DCS tone sequence for access
D. All of these choices are correct

5. _____ T2B05 What determines the amount of deviation of an FM signal?
A. Both the frequency and amplitude of the modulating signal
B. The frequency of the modulating signal
C. The amplitude of the modulating signal
D. The relative phase of the modulating signal and the carrier

6. _____ T2B06 What happens when the deviation of an FM transmitter is increased?
A. Its signal occupies more bandwidth
B. Its output power increases
C. Its output power and bandwidth increases
D. Asymmetric modulation occurs

7. _____ T2B07 What should you do if you receive a report that your station's transmissions are causing splatter or interference on nearby frequencies?
A. Increase transmit power
B. Change mode of transmission
C. Report the interference to the equipment manufacturer
D. Check your transmitter for off-frequency operation or spurious emissions

8. _____ T2B08 What is the proper course of action if your station's transmission unintentionally interferes with another station?
A. Rotate your antenna slightly
B. Properly identify your transmission and move to a different frequency
C. Increase power
D. Change antenna polarization

9. _____ T2B09 [97.119(b)(2)] Which of the following methods is encouraged by the FCC when identifying your station when using phone?
A. Use of a phonetic alphabet
B. Send your call sign in CW as well as voice
C. Repeat your call sign three times
D. Increase your signal to full power when identifying

10. _____ T2B10 What is the "Q" signal used to indicate that you are receiving interference from other stations?
A. QRM
B. QRN
C. QTH
D. QSB

11. _____ T2B11 What is the "Q" signal used to indicate that you are changing frequency?
A. QRU
B. QSY
C. QSL
D. QRZ

Lesson 8 Study Sheet

✓ CTCSS stands for Continuous Tone-Codes Squelch System. It is the use of a sub-audible tone transmitted with normal voice audio to open the squelch of a receiver.

✓ The deviation of an FM signal is determined by the amplitude of the modulating signal.

✓ When the deviation of an FM transmitter is increased its signal occupies more bandwidth.

✓ QRM is a "Q" signal meaning the station is receiving interference from other stations.

✓ QSY is the "Q" signal used to indicate that you are changing frequency.

Notes:

Lesson 9
The T2C Section
Public Service Emergency and Non-emergency Operations

T2C01 [97.103(a)] What set of rules applies to proper operation of your station when using amateur radio at the request of public service officials?
Answer: *FCC Rules.*

The FCC rules always apply to proper amateur operation. Always.

T2C02 [97.113 and FCC Public Notice DA 09-2259] Who must submit the request for a temporary waiver of Part 97.113 to allow amateur radio operators to provide communications on behalf of their employers during a government sponsored disaster drill?
Answer: *The government agency sponsoring the event.*

This applies if the company that you work for is going to take part in a government sponsored disaster drill and you are expected to provide amateur radio communications on behalf of that company. To provide amateur radio communications for the company you work for, a waiver is needed from the FCC. It is the responsibility of the government agency sponsoring the drill to get that waiver.

T2C03 [97.113] When is it legal for an amateur licensee to provide communications on behalf of their employer during a government sponsored disaster drill or exercise?
Answer: *Only when the FCC has granted a government-requested waiver.*

This question is building on the previous question. You need to have a waiver from the FCC to provide amateur radio communications on behalf of your employer during a government sponsored drill.

T2C04 What do RACES and ARES have in common?
Answer: *Both organizations may provide communications during emergencies.*

RACES stands for Radio Amateur Civil Emergency Service. RACES are hams who have registered their stations with a civil defense organization. RACES primary role is civil defense but can be used in natural disasters and other emergencies as well. ARES stands for Amateur Radio Emergency Service and is organized by the ARRL. ARES is a membership of amateur operators who provide their skills and equipment to support disaster relief communications.

T2C05 [97.3(a)(37), 97.407] What is the Radio Amateur Civil Emergency Service?
Answer: *A radio service using amateur stations for emergency management or civil defense communications.*

The Radio Amateur Civil Emergency Service, also known as RACES.

T2C06 Which of the following is common practice during net operations to get the immediate attention of the net control station when reporting an emergency?
Answer: *Begin your transmission with "Priority" or "Emergency" followed by your call sign.*

To get the attention of the net control operator, the first word you send needs to be "Priority" or "Emergency." The next word needs to be your call sign. Once the net control station acknowledges you, then you can pass your emergency information.

T2C07 What should you do to minimize disruptions to an emergency traffic net once you have checked in?
Answer: *Do not transmit on the net frequency until asked to do so by the net control station.*

Let the net control station run the show. When the net control station is dealing with an emergency, it is not the time to get chatty. After you check into the net, the net control station will run down the list of checked-in stations. When the net calls on you then you may pass your information.

T2C08 What is usually considered to be the most important job of an amateur operator when handling emergency traffic messages?
Answer: *Passing messages exactly as written, spoken or as received.*

Pass the message exactly as received! They mean exactly down to the number of words! Your job is not to guess or interpret information. Your job is to pass the message to the person whose job it is to interpret the information received.

T2C09 [97.403] When may an amateur station use any means of radio communications at its disposal for essential communications in connection with immediate safety of human life and protection of property?
Answer: *When normal communications systems are not available.*

When normal communication systems are not available, and it is a bona fide life or death emergency, the Part 97 rules permit you to do whatever it takes to pass information to assist those in danger.

T2C10 What is the preamble in a formal traffic message?
Answer: *The information needed to track the message as it passes through the amateur radio traffic handling system.*

The preamble in a formal message contains all the necessary information about who sent the message and where it's going. The preamble includes information such as a unique identification number for that message, who sent the message and their location, the word count (or check), and any special handling instructions.

T2C11 What is meant by the term "check" in reference to a formal traffic message?
Answer: *The check is a count of the number of words or word equivalents in the text portion of the message.*

Remember the most important job of an amateur operator when dealing with emergency communications is to pass the message exactly as received. One of the ways that level of exactness is achieved is the check. The check is the number of words or word equivalents in a message. If the number you copied is different from the check, start over.

Quiz #9

(Find the answers on page 256)

1. _____ T2C01 [97.103(a)] What set of rules applies to proper operation of your station when using amateur radio at the request of public service officials?
A. RACES Rules
B. ARES Rules
C. FCC Rules
D. FEMA Rules

2. _____ T2C02 [97.113 and FCC Public Notice DA 09-2259] Who must submit the request for a temporary waiver of Part 97.113 to allow amateur radio operators to provide communications on behalf of their employers during a government sponsored disaster drill?
A. Each amateur participating in the drill
B. Any employer participating in the drill
C. The local American Red Cross Chapter
D. The government agency sponsoring the event

3. _____ T2C03 [97.113] When is it legal for an amateur licensee to provide communications on behalf of their employer during a government sponsored disaster drill or exercise?
A. Whenever the employer is a not-for-profit organization
B. Whenever there is a temporary need for the employer's business continuity plan
C. Only when the FCC has granted a government-requested waiver
D. Only when the amateur is not receiving compensation from his employer for the activity

4. _____ T2C04 What do RACES and ARES have in common?
A. They represent the two largest ham clubs in the United States
B. Both organizations broadcast road and weather traffic information
C. Neither may handle emergency traffic supporting public service agencies
D. Both organizations may provide communications during emergencies

5. _____ T2C05 [97.3(a)(37), 97.407] What is the Radio Amateur Civil Emergency Service?
A. An emergency radio service organized by amateur operators
B. A radio service using amateur stations for emergency management or civil defense communications
C. A radio service organized to provide communications at civic events
D. A radio service organized by amateur operators to assist non-military persons

6. _____ T2C06 Which of the following is common practice during net operations to get the immediate attention of the net control station when reporting an emergency?
A. Repeat the words SOS three times followed by the call sign of the reporting station
B. Press the push-to-talk button three times
C. Begin your transmission with "Priority" or "Emergency" followed by your call sign
D. Play a pre-recorded emergency alert tone followed by your call sign

7. _____ T2C07 What should you do to minimize disruptions to an emergency traffic net once you have checked in?
A. Whenever the net frequency is quiet, announce your call sign and location
B. Move 5 kHz away from the net's frequency and use high power to ask other hams to keep clear of the net frequency
C. Do not transmit on the net frequency until asked to do so by the net control station
D. Wait until the net frequency is quiet, then ask for any emergency traffic for your area

8. _____ T2C08 What is usually considered to be the most important job of an amateur operator when handling emergency traffic messages?
A. Passing messages exactly as written, spoken or as received
B. Estimating the number of people affected by the disaster
C. Communicating messages to the news media for broadcast outside the disaster area
D. Broadcasting emergency information to the general public

9. _____ T2C09 [97.403] When may an amateur station use any means of radio communications at its disposal for essential communications in connection with immediate safety of human life and protection of property?
A. Only when FEMA authorizes it by declaring an emergency
B. When normal communications systems are not available
C. Only when RACES authorizes it by declaring an emergency
D. Only when authorized by the local MARS program director

10. _____ T2C10 What is the preamble in a formal traffic message?
A. The first paragraph of the message text
B. The message number
C. The priority handling indicator for the message
D. The information needed to track the message as it passes through the amateur radio traffic handling system

11. _____ T2C11 What is meant by the term "check" in reference to a formal traffic message?
A. The check is a count of the number of words or word equivalents in the text portion of the message
B. The check is the value of a money order attached to the message
C. The check is a list of stations that have relayed the message
D. The check is a box on the message form that tells you the message was received

Lesson 9 Study Sheet

✓ RACES stands for Radio Amateur Civil Emergency Service.

✓ ARES stands for Amateur Radio Emergency Service.

✓ The most important job of an amateur operator when handling emergency traffic messages is to pass messages exactly as written, spoken or as received.

✓ The check is a count of the number of words, or word equivalents, in the text portion of the message.

Notes:

Lesson 10
The T3A Section
Radio Wave Characteristics

T3A01 What should you do if another operator reports that your station's 2 meter signals were strong just a moment ago, but now they are weak or distorted?
Answer: *Try moving a few feet, as random reflections may be causing multi-path distortion.*

Have you ever pulled up to a stop light at a downtown intersection and noticed your FM radio picking up a little static which goes away as you creep up to the light? Reflections and distortions can create small pockets of low reception. Moving a few feet will usually solve the problem.

T3A02 Why are UHF signals often more effective from inside buildings than VHF signals?
Answer: *The shorter wavelength allows them to more easily penetrate the structure of buildings.*

Smaller wavelengths can penetrate solid objects better than longer wavelengths. UHF signals have a much smaller wavelength than VHF. For instance, the 2 meter band in the VHF range has a wavelength of roughly 2 meters. The UHF bands are looking at wavelengths like 70 cm, 23 cm, etc.

T3A03 What antenna polarization is normally used for long-distance weak-signal CW and SSB contacts using the VHF and UHF bands?
Answer: *Horizontal.*

For the purposes of the exam, there are only two antenna polarizations you need to worry about: horizontal and vertical. These two types of polarization generally are associated with how the antenna is positioned relative to the Earth. Vertical antennas that are perpendicular to the ground will generally have a vertical polarization. Antennas such as dipoles and Yagis whose elements are horizontal to the ground will have a horizontal polarization. For VHF and UHF CW and SSB signals, horizontally polarized signals experience less ground loss than vertically polarized signals and thus travel farther. A way to help remember this is to imagine throwing a Frisbee. If you throw the Frisbee parallel to the ground it will travel farther than if you throw it vertically like a tomahawk.

T3A04 What can happen if the antennas at opposite ends of a VHF or UHF line of sight radio link are not using the same polarization?
Answer: *Signals could be significantly weaker.*

Horizontally polarized antennas pick up horizontally polarized signals better than vertically polarized antennas, and visa versa. If the polarization of the signal is different from the polarization of the antenna receiving the signal, much of the signal will not be received resulting in significantly weaker reception.

T3A05 When using a directional antenna, how might your station be able to access a distant repeater if buildings or obstructions are blocking the direct line of sight path?
Answer: *Try to find a path that reflects signals to the repeater.*

VHF and UHF communications are sometimes called line of sight communications implying that you can communicate with stations that your signal has a clear path for the signal to travel. However, VHF and UHF signals can sometimes be bounced off large objects. If a mountain or a large building is blocking the line of sight between you and the station you are trying to communicate with, try pointing your directional antenna at the side of another building, or mountain, and reflect the signal to the receiving station.

T3A06 What term is commonly used to describe the rapid fluttering sound sometimes heard from mobile stations that are moving while transmitting?
Answer: *Picket fencing.*

Picket fencing is a very good description on how this sounds. When you are transmitting in a moving vehicle, your signal is constantly bouncing off of different objects as you pass them by causing a fluttering sound in the receiving station. Imagine the sound you make when you talk into a spinning fan.

T3A07 What type of wave carries radio signals between transmitting and receiving stations?
Answer: *Electromagnetic.*

This is an important one to know for general purposes. Electromagnetic waves and how to get them to carry information to where you want them to go is the soul of amateur radio. Electromagnetic waves consist of an electric and magnetic field, thus their name.

T3A08 What is the cause of irregular fading of signals from distant stations during times of generally good reception?
Answer: *Random combining of signals arriving via different path lengths.*

Radio waves can do funny things when they run into each other. One of the things that can happen is they can cancel each other out when they collide. You can see this in a pool of water. If two people are in the pool and push a wave at each other, when the waves meet each other in the middle of the pool the two waves will cancel each other out.

T3A09 Which of the following is a common effect of "skip" reflections between the Earth and the ionosphere?
Answer: *The polarization of the original signal is randomized.*

Radio signals can be bounced off the ionosphere under certain conditions. This is especially true for the HF bands allowing for very long distance communications on those bands. For this question you need to know that regardless of the polarization of the antenna sending the signal, once the signal is bounced off the ionosphere the polarization of the signal can change.

T3A10 What may occur if VHF or UHF data signals propagate over multiple paths?
Answer: *Error rates are likely to increase.*

Data signals suffer from the same issues as voice communications when signals run into each other over multiple paths. The same way you may have trouble understanding speech in faded or distorted signals, your computer may not understand all the data it is receiving. The result is an increased error rate in the received information.

T3A11 Which part of the atmosphere enables the propagation of radio signals around the world?
Answer: *The ionosphere.*

Bouncing signals off the ionosphere is responsible for the majority of DX (long distance) contacts in amateur radio. The HF bands are the best bands for ionospheric skip, but sometimes conditions allow skip with the 2 meter and 6 meter bands. Depending on the frequency, a single bounce off the ionosphere can reach distances up to 2,500 miles. Multiple bounces between the earth and ionosphere are possible allowing signals to reach anywhere in the world!

Quiz #10

(Find the answers on page 256)

1. _____ T3A01 What should you do if another operator reports that your station's 2 meter signals were strong just a moment ago, but now they are weak or distorted?
A. Change the batteries in your radio to a different type
B. Turn on the CTCSS tone
C. Ask the other operator to adjust his squelch control
D. Try moving a few feet, as random reflections may be causing multi-path distortion

2. _____ T3A02 Why are UHF signals often more effective from inside buildings than VHF signals?
A. VHF signals lose power faster over distance
B. The shorter wavelength allows them to more easily penetrate the structure of buildings
C. This is incorrect; VHF works better than UHF inside buildings
D. UHF antennas are more efficient than VHF antennas

3. _____ T3A03 What antenna polarization is normally used for long-distance weak-signal CW and SSB contacts using the VHF and UHF bands?
A. Right-hand circular
B. Left-hand circular
C. Horizontal
D. Vertical

4. _____ T3A04 What can happen if the antennas at opposite ends of a VHF or UHF line of sight radio link are not using the same polarization?
A. The modulation sidebands might become inverted
B. Signals could be significantly weaker
C. Signals have an echo effect on voices
D. Nothing significant will happen

5. _____ T3A05 When using a directional antenna, how might your station be able to access a distant repeater if buildings or obstructions are blocking the direct line of sight path?
A. Change from vertical to horizontal polarization
B. Try to find a path that reflects signals to the repeater
C. Try the long path
D. Increase the antenna SWR

6. _____ T3A06 What term is commonly used to describe the rapid fluttering sound sometimes heard from mobile stations that are moving while transmitting?
A. Flip-flopping
B. Picket fencing
C. Frequency shifting
D. Pulsing

7. _____ T3A07 What type of wave carries radio signals between transmitting and receiving stations?
A. Electromagnetic
B. Electrostatic
C. Surface acoustic
D. Magnetostrictive

8. _____ T3A08 What is the cause of irregular fading of signals from distant stations during times of generally good reception?
A. Absorption of signals by the "D" layer of the ionosphere
B. Absorption of signals by the "E" layer of the ionosphere
C. Random combining of signals arriving via different path lengths
D. Intermodulation distortion in the local receiver

9. _____ T3A09 Which of the following is a common effect of "skip" reflections between the Earth and the ionosphere?
A. The sidebands become reversed at each reflection
B. The polarization of the original signal is randomized
C. The apparent frequency of the received signal is shifted by a random amount
D. Signals at frequencies above 30 MHz become stronger with each reflection

10. _____ T3A10 What may occur if VHF or UHF data signals propagate over multiple paths?
A. Transmission rates can be increased by a factor equal to the number of separate paths observed
B. Transmission rates must be decreased by a factor equal to the number of separate paths observed
C. No significant changes will occur if the signals are transmitting using FM
D. Error rates are likely to increase

11. _____ T3A11 Which part of the atmosphere enables the propagation of radio signals around the world?
A. The stratosphere
B. The troposphere
C. The ionosphere
D. The magnetosphere

Lesson 10 Study Sheet

✓ Picket fencing is a term commonly used to describe the rapid fluttering sound sometimes heard from mobile stations that are moving while transmitting

✓ Electromagnetic waves carry radio signals between transmitting and receiving stations.

✓ The ionosphere is the part of the atmosphere that enables the propagation of radio signals around the world.

Notes:

Lesson 11
The T3B Section
Radio and Electromagnetic Wave Properties

T3B01 What is the name for the distance a radio wave travels during one complete cycle?
Answer: *Wavelength.*

Wavelength is exactly what it says it is, the length of a wave. The wavelength of a radio signal is determined by the distance of one complete wave cycle. The wavelength of frequencies used by amateurs can be as short as 23 cm to as long as 160 meters.

T3B02 What term describes the number of times per second that an alternating current reverses direction?
Answer: *Frequency.*

For the purposes of the exam, frequency is the number of times per second something happens. We have talked about frequency in terms of radio waves, but frequency applies to the cycle of alternating current (AC) as well. Alternating current is the change in direction of the flow of electricity in a circuit from one direction to the opposite direction. The number of times in a second this flow of electricity reverses itself is its frequency.

T3B03 What are the two components of a radio wave?
Answer: *Electric and magnetic fields.*

Radio waves are electromagnetic waves. The two components of an electromagnet wave are electric fields and magnetic fields.

T3B04 How fast does a radio wave travel through free space?
Answer: *At the speed of light.*

Radio waves travel at the speed of light, or 300,000,000 meters per second. That's as fast as anything can go!

T3B05 How does the wavelength of a radio wave relate to its frequency?
Answer: *The wavelength gets shorter as the frequency increases.*

There is an inverse relationship between wavelength and frequency. The higher the frequency, the shorter the wavelength. The highest frequencies amateurs can work are in the 23 cm band and have a short wavelength of 23 cm. The lowest frequencies which are within amateur privileges are in the 160 meter band and have a wavelength of 160 meters.

T3B06 What is the formula for converting frequency to wavelength in meters?
Answer: *Wavelength in meters equals 300 divided by frequency in megahertz.*

You should memorize this formula:
Wavelength (in meters) = 300/Frequency (in MHz)

T3B07 What property of radio waves is often used to identify the different frequency bands?
Answer: *The approximate wavelength.*

The radio wave's wavelength is used to categorize them in the various bands. For instance, the 40 meter band, 10 meter band, 2 meter band, 70 cm band, etc.

T3B08 What are the frequency limits of the VHF spectrum?
Answer: *30 to 300 MHz.*

The privileges you will have as a Technician Class licensee are in the HF, VHF, and UHF ranges. The UHF range have the highest frequencies, the HF range has the lowest frequencies, and the VHF range is right there in the middle. The limits of the VHF spectrum is 30 MHz to 300 MHz. You need to memorize this for the exam.

T3B09 What are the frequency limits of the UHF spectrum?
Answer: *300 to 3000 MHz.*

The UHF spectrum is the highest frequency range in the amateur privileges. The UHF spectrum ranges from 300 MHz to 3000 MHz.

T3B10 What frequency range is referred to as HF?
Answer: *3 to 30 MHz.*

HF is the lowest spectrum that a Technician Class operator has privileges in, but not the lowest for General and Extra Class licensees. The HF spectrum ranges from 3 to 30 MHz.

T3B11 What is the approximate velocity of a radio wave as it travels through free space?
Answer: *300,000,000 meters per second.*

This is the speed of light. Radio waves travel at 300,000,000 meters per second!

Quiz #11

(Find the answers on page 256)

1. _____ T3B01 What is the name for the distance a radio wave travels during one complete cycle?
A. Wave speed
B. Waveform
C. Wavelength
D. Wave spread

2. _____ T3B02 What term describes the number of times per second that an alternating current reverses direction?
A. Pulse rate
B. Speed
C. Wavelength
D. Frequency

3. _____ T3B03 What are the two components of a radio wave?
A. AC and DC
B. Voltage and current
C. Electric and magnetic fields
D. Ionizing and non-ionizing radiation

4. _____ T3B04 How fast does a radio wave travel through free space?
A. At the speed of light
B. At the speed of sound
C. Its speed is inversely proportional to its wavelength
D. Its speed increases as the frequency increases

5. _____ T3B05 How does the wavelength of a radio wave relate to its frequency?
A. The wavelength gets longer as the frequency increases
B. The wavelength gets shorter as the frequency increases
C. There is no relationship between wavelength and frequency
D. The wavelength depends on the bandwidth of the signal

6. _____ T3B06 What is the formula for converting frequency to wavelength in meters?
A. Wavelength in meters equals frequency in hertz multiplied by 300
B. Wavelength in meters equals frequency in hertz divided by 300
C. Wavelength in meters equals frequency in megahertz divided by 300
D. Wavelength in meters equals 300 divided by frequency in megahertz

7. _____ T3B07 What property of radio waves is often used to identify the different frequency bands?
A. The approximate wavelength
B. The magnetic intensity of waves
C. The time it takes for waves to travel one mile
D. The voltage standing wave ratio of waves

8. _____T3B08 What are the frequency limits of the VHF spectrum?
A. 30 to 300 kHz
B. 30 to 300 MHz
C. 300 to 3000 kHz
D. 300 to 3000 MHz

9. _____ T3B09 What are the frequency limits of the UHF spectrum?
A. 30 to 300 kHz
B. 30 to 300 MHz
C. 300 to 3000 kHz
D. 300 to 3000 MHz

10. _____ T3B10 What frequency range is referred to as HF?
A. 300 to 3000 MHz
B. 30 to 300 MHz
C. 3 to 30 MHz
D. 300 to 3000 kHz

11. _____ T3B11 What is the approximate velocity of a radio wave as it travels through free space?
A. 3000 kilometers per second
B. 300,000,000 meters per second
C. 300,000 miles per hour
D. 186,000 miles per hour

Lesson 11 Study Sheet

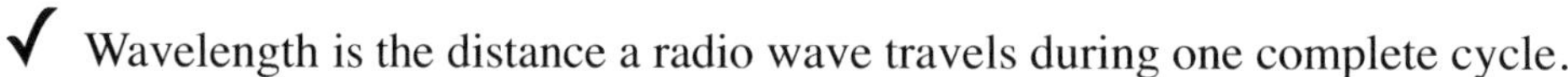

✓ Wavelength is the distance a radio wave travels during one complete cycle.

✓ Frequency for alternating current is the number of times per second that the current reverses direction.

✓ A radio wave is an electromagnetic wave. The components of electromagnetic waves are electric and magnetic fields.

✓ Radio waves travel at the speed of light, or 300,000,000 meters per second.

✓ Wavelength (in meters) = 300/Frequency (in MHz)

✓ The HF spectrum ranges from 3 to 30 MHz.

✓ The VHF spectrum ranges from 30 MHz to 300 MHz.

✓ The UHF spectrum ranges from 300 MHz to 3000 MHz.

Notes:

Lesson 12
The T3C Section
Propagation Modes

T3C01 Why are "direct" (not via a repeater) UHF signals rarely heard from stations outside your local coverage area?
Answer: *UHF signals are usually not reflected by the ionosphere.*

UHF communications are line of sight communications. The chances for ionospheric conditions to support bending back UHF signals to go beyond the line of sight is extremely remote. Unless you are able to connect through a repeater, UHF signals stick to the local coverage area.

T3C02 Which of the following might be happening when VHF signals are being received from long distances?
Answer: *Signals are being refracted from a sporadic E layer.*

Occasionally, or sporadically, the sun will ionize patches in the E layer of the ionosphere. If these patches are sufficiently charged, they are capable of reflecting VHF signals back to the Earth. Usually these patches are discovered by surprise after you realize the location of the amateur on the other end is over a thousand miles away! Sporadic E can also be found in the 10 meter and 6 meter bands.

T3C03 What is a characteristic of VHF signals received via auroral reflection?
Answer: *The signals exhibit rapid fluctuations of strength and often sound distorted.*

Yes, you can bounce VHF signals off of the Northern Lights! Auroras are charged particles cascading through the ionosphere. The uneven flow of these charged particles not only gives the Norther Lights their wavelike appearance, they also cause fluctuations and distortions to the VHF and UHF radio signals that bounce off them.

T3C04 Which of the following propagation types is most commonly associated with occasional strong over-the-horizon signals on the 10, 6, and 2 meter bands?
Answer: *Sporadic E.*

When you experience occasional, or sporadic, strong over the horizon communications on the 10 meter, 6 meter, and 2 meter bands, it is probably sporadic E.

T3C05 What is meant by the term "knife-edge" propagation?
Answer: *Signals are partially refracted around solid objects exhibiting sharp edges.*

VHF and UHF radio signals are generally blocked by objects. However, occasionally when VHF and UHF signals pass the sharp edge of an obstacle they can refract, or bend, into what would otherwise be the signal's shadow behind the object. The key word in both the question and the answer is "edge." Remember, knife edges have sharp edges.

T3C06 What mode is responsible for allowing over-the-horizon VHF and UHF communications to ranges of approximately 300 miles on a regular basis?
Answer: *Tropospheric scatter.*

Certain weather conditions in the troposphere will allow VHF and UHF signals to travel over the horizon. Storm fronts or large temperature changes often support tropospheric scatter and your signal can reach ranges up to 300 miles.

T3C07 What band is best suited to communicating via meteor scatter?
Answer: 6 meters.

As a meteor burns up in the atmosphere it leaves an ionized particle trail behind it. You can bounce signals off these ionized trails. Of all the bands, 6 meters works the best.

T3C08 What causes "tropospheric ducting"?
Answer: *Temperature inversions in the atmosphere.*

These temperature inversions create ducts in the troposphere which can carry VHF and UHF signals incredibly long distances.

T3C09 What is generally the best time for long-distance 10 meter band propagation?
Answer: *During daylight hours.*

10 meters works best with a highly charged ionosphere. The sun is what is responsible for the vast majority of the ionization of the ionosphere. Therefore it only makes sense that the ionosphere will be most charged during the day and you will get the longest distance out of 10 meters.

T3C10 What is the radio horizon?
Answer: *The distance at which radio signals between two points are effectively blocked by the curvature of the Earth.*

VHF and UHF communications are generally line of sight. Assuming these signals are not blocked by a mountain or other obstacle, eventually the curvature of the Earth will block the signal. The point at which the curvature of the Earth blocks radio signals from a station is called the radio horizon.

T3C11 Why do VHF and UHF radio signals usually travel somewhat farther than the visual line of sight distance between two stations?
Answer: *The Earth seems less curved to radio waves than to light.*

Radio waves bend, or diffract, slightly more than light. This allows radio waves to travel a little bit farther than the visual horizon. This bending allows radio waves to follow the surface of the Earth a little farther than what is visible.

Quiz #12

(Find the answers on page 256)

1. _____ T3C01 Why are "direct" (not via a repeater) UHF signals rarely heard from stations outside your local coverage area?
A. They are too weak to go very far
B. FCC regulations prohibit them from going more than 50 miles
C. UHF signals are usually not reflected by the ionosphere
D. They collide with trees and shrubbery and fade out

2. _____ T3C02 Which of the following might be happening when VHF signals are being received from long distances?
A. Signals are being reflected from outer space
B. Signals are arriving by sub-surface ducting
C. Signals are being reflected by lightning storms in your area
D. Signals are being refracted from a sporadic E layer

3. _____ T3C03 What is a characteristic of VHF signals received via auroral reflection?
A. Signals from distances of 10,000 or more miles are common
B. The signals exhibit rapid fluctuations of strength and often sound distorted
C. These types of signals occur only during winter nighttime hours
D. These types of signals are generally strongest when your antenna is aimed to the south (for stations in the Northern Hemisphere)

4. _____ T3C04 Which of the following propagation types is most commonly associated with occasional strong over-the-horizon signals on the 10, 6, and 2 meter bands?
A. Backscatter
B. Sporadic E
C. D layer absorption
D. Gray-line propagation

5. _____ T3C05 What is meant by the term "knife-edge" propagation?
A. Signals are reflected back toward the originating station at acute angles
B. Signals are sliced into several discrete beams and arrive via different paths
C. Signals are partially refracted around solid objects exhibiting sharp edges
D. Signals propagated close to the band edge exhibiting a sharp cutoff

6. _____ T3C06 What mode is responsible for allowing over-the-horizon VHF and UHF communications to ranges of approximately 300 miles on a regular basis?
A. Tropospheric scatter
B. D layer refraction
C. F2 layer refraction
D. Faraday rotation

7. _____ T3C07 What band is best suited to communicating via meteor scatter?
A. 10 meters
B. 6 meters
C. 2 meters
D. 70 cm

8. _____ T3C08 What causes "tropospheric ducting"?
A. Discharges of lightning during electrical storms
B. Sunspots and solar flares
C. Updrafts from hurricanes and tornadoes
D. Temperature inversions in the atmosphere

9. _____ T3C09 What is generally the best time for long-distance 10 meter band propagation?
A. During daylight hours
B. During nighttime hours
C. When there are coronal mass ejections
D. Whenever the solar flux is low

10. _____ T3C10 What is the radio horizon?
A. The distance at which radio signals between two points are effectively blocked by the curvature of the Earth
B. The distance from the ground to a horizontally mounted antenna
C. The farthest point you can see when standing at the base of your antenna tower
D. The shortest distance between two points on the Earth's surface

11. _____ T3C11 Why do VHF and UHF radio signals usually travel somewhat farther than the visual line of sight distance between two stations?
A. Radio signals move somewhat faster than the speed of light
B. Radio waves are not blocked by dust particles
C. The Earth seems less curved to radio waves than to light
D. Radio waves are blocked by dust particles

Lesson 12 Study Sheet

✔ Sporadic E are occasionally occurring ionized patches in the E layer of the ionosphere. Sporadic E can support reflecting 10 meter, 6 meter, and 2 meter band signals permitting long distance communications.

✔ "Knife-edge" propagation deals with signals that are partially refracted around solid objects exhibiting sharp edges.

✔ Tropospheric scatter allows for VHF and UHF contacts to regularly reach distances of 300 miles.

✔ Frequencies within the 6 meter band are best suited to communicating via meteor scatter.

✔ The best time for 10 meter long-distance propagation is during the day.

Notes:

Lesson 13
The T4A Section
Station Setup

T4A01 Which of the following is true concerning the microphone connectors on amateur transceivers?
Answer: *Some connectors include push-to-talk and voltages for powering the microphone.*

Not all microphones are made the same. Therefore, it is safe to say not all microphone connectors are made the same. Some microphones have push to talk functions, some require power, and some even have speakers built in. Depending on the microphone features and capabilities of the transceiver, different brands of transceivers will have different connectors with different wiring configurations for the microphone.

T4A02 What could be used in place of a regular speaker to help you copy signals in a noisy area?
Answer: *A set of headphones.*

A good set of headphones is an absolute must. Not only does it help you copy signals in a noisy area, they can help you hear weak signals better.

T4A03 Which is a good reason to use a regulated power supply for communications equipment?
Answer: *It prevents voltage fluctuations from reaching sensitive circuits.*

A regulated power supply keeps the power going to your equipment in a nice and steady flow and prevents fluctuations in the voltage. Voltage fluctuations can damage your equipment. A regulated power supply helps protect your equipment.

T4A04 Where must a filter be installed to reduce harmonic emissions?
Answer: *Between the transmitter and the antenna.*

The key word in the question is "emissions." Harmonic emission is something your transmitter produces. Of the possible answers, the only logical place to filter emissions from the transmitter is between the transmitter and the antenna. If your transmitter is causing interference to other electronic devices, try putting a filter between your transmitter and antenna.

T4A05 What type of filter should be connected to a TV receiver as the first step in trying to prevent RF overload from a nearby 2 meter transmitter?
Answer: *Band-reject filter.*

You do not want any interference from your 2 meter transmitter messing with your, or your neighbor's, TV. You want to use a filter that will reject all interference without filtering out the television signals. A band-reject filter just blocks or rejects a specific band or range of frequencies, like unwanted 2 meter interference.

T4A06 Which of the following would be connected between a transceiver and computer in a packet radio station?
Answer: *Terminal node controller.*

A terminal node controller, or TNC, is used as the bridge between the radio and computer in a packet station. A TNC converts the radio signals received into a digital form the computer can understand. When sending transmissions it will convert the digital information back into audio form for the transmitter to send.

T4A07 How is the computer's sound card used when conducting digital communications using a computer?
Answer: *The sound card provides audio to the microphone input and converts received audio to digital form.*

The use of the computer's sound card to conduct digital operations is becoming increasingly popular. The computer's sound card can preform the same type of functions as a TNC does in packet operation. The sound card can convert audio to digital and digital to audio.

T4A08 Which type of conductor is best to use for RF grounding?
Answer: *Flat strap.*

A flat copper strap is the best for RF grounding. The wider and shorter the better (within reason). This type of conductor has the least amount of impedance of the possible answers and works best for RF grounding.

T4A09 Which would you use to reduce RF current flowing on the shield of an audio cable?
Answer: *Ferrite choke.*

RF current flowing on an unshielded audio cable can cause a lot of trouble. You want to avoid any RF current flowing on the shield of an audio cable. You don't want to filter the RF current, you want to choke it out. A ferrite choke will work.

T4A10 What is the source of a high-pitched whine that varies with engine speed in a mobile transceiver's receive audio?
Answer: *The alternator.*

If you hear a whine in your mobile transceiver's audio that increases and decreases with the speed of your engine, it is most likely caused by the alternator. In most vehicles, the alternator is driven directly by the engine and produces the electricity that powers the vehicle and recharges the battery. If you are having problems with the alternator producing a whine in your mobile transceiver, check to make sure the transceiver is grounded well and you might try hooking the negative connection of your transceiver directly to the vehicle's battery.

T4A11 Where should a mobile transceiver's power negative connection be made?
Answer: *At the battery or engine block ground strap.*

For proper performance, you want the best connection possible for your mobile transceiver's negative connection. Simply hooking the negative connection to any piece of metal on the vehicle may not allow a good solid connection or steady flow of power to your transceiver. A good solid connection to the vehicle's battery or the engine block's ground strap is the most reliable way to ensure a steady flow of power to your transceiver.

Quiz #13

(Find the answers on page 256)

1. _____ T4A01 Which of the following is true concerning the microphone connectors on amateur transceivers?
A. All transceivers use the same microphone connector type
B. Some connectors include push-to-talk and voltages for powering the microphone
C. All transceivers using the same connector type are wired identically
D. Un-keyed connectors allow any microphone to be connected

2. _____ T4A02 What could be used in place of a regular speaker to help you copy signals in a noisy area?
A. A video display
B. A low pass filter
C. A set of headphones
D. A boom microphone

3. _____ T4A03 Which is a good reason to use a regulated power supply for communications equipment?
A. It prevents voltage fluctuations from reaching sensitive circuits
B. A regulated power supply has FCC approval
C. A fuse or circuit breaker regulates the power
D. Power consumption is independent of load

4. _____ T4A04 Where must a filter be installed to reduce harmonic emissions?
A. Between the transmitter and the antenna
B. Between the receiver and the transmitter
C. At the station power supply
D. At the microphone

5. _____ T4A05 What type of filter should be connected to a TV receiver as the first step in trying to prevent RF overload from a nearby 2 meter transmitter?
A. Low-pass filter
B. High-pass filter
C. Band-pass filter
D. Band-reject filter

6. _____ T4A06 Which of the following would be connected between a transceiver and computer in a packet radio station?
A. Transmatch
B. Mixer
C. Terminal node controller
D. Antenna

7. _____ T4A07 How is the computer's sound card used when conducting digital communications using a computer?
A. The sound card communicates between the computer CPU and the video display
B. The sound card records the audio frequency for video display
C. The sound card provides audio to the microphone input and converts received audio to digital form
D. All of these choices are correct

8. _____ T4A08 Which type of conductor is best to use for RF grounding?
A. Round stranded wire
B. Round copper-clad steel wire
C. Twisted-pair cable
D. Flat strap

9. _____ T4A09 Which would you use to reduce RF current flowing on the shield of an audio cable?
A. Band-pass filter
B. Low-pass filter
C. Preamplifier
D. Ferrite choke

10. _____ T4A10 What is the source of a high-pitched whine that varies with engine speed in a mobile transceiver's receive audio?
A. The ignition system
B. The alternator
C. The electric fuel pump
D. Anti-lock braking system controllers

11. _____ T4A11 Where should a mobile transceiver's power negative connection be made?
A. At the battery or engine block ground strap
B. At the antenna mount
C. To any metal part of the vehicle
D. Through the transceiver's mounting bracket

Lesson 13 Study Sheet

✓ A band-reject filter should be connected to a TV receiver as the first step in trying to prevent RF overload from a nearby 2 meter transmitter.

✓ A terminal node controller (TNC) is connected between a transceiver and computer in a packet radio station.

✓ A ferrite choke is used to reduce RF current flowing on the shield of an audio cable.

Notes:

Lesson 14
The T4B Section
Operating Controls

T4B01 What may happen if a transmitter is operated with the microphone gain set too high?
Answer: *The output signal might become distorted.*

The microphone gain on your transmitter adjusts the dynamics of your microphone. If the gain is set too high it can cause over modulation and distortion in signals. You can be sure that another ham will let you know if you are operating with the microphone gain set too high.

T4B02 Which of the following can be used to enter the operating frequency on a modern transceiver?
Answer: *The keypad or VFO knob.*

VFO stands for variable frequency oscillator and is the circuit in your transceiver that controls what frequency you are operating on. You can enter the operating frequency on a modern transceiver by turning the VFO's knob or just simply entering the frequency into the keypad. Don't over think the question!

T4B03 What is the purpose of the squelch control on a transceiver?
Answer: *To mute receiver output noise when no signal is being received.*

The squelch on a receiver is what prevents you from sitting there listening to static on a frequency when nobody else is transmitting. The squelch keeps the receiver muted until the receiver picks up a signal.

T4B04 What is a way to enable quick access to a favorite frequency on your transceiver?
Answer: *Store the frequency in a memory channel.*

Most modern transceivers, especially handheld transceivers (HTs), are capable of storing frequencies in memory. A quick way to access a favorite frequency is to store it in a memory channel.

T4B05 Which of the following would reduce ignition interference to a receiver?
Answer: *Turn on the noise blanker.*

The noise blanker circuit in your mobile receiver basically blanks out noise. The spark plugs in your vehicle's engine blast a wide range of frequencies every time they fire. Simply trying to tune away from the noise usually will not work. Switching on the noise blanker will help eliminate this ignition noise

T4B06 Which of the following controls could be used if the voice pitch of a single-sideband signal seems too high or low?
Answer: *The receiver RIT or clarifier.*

RIT stands for receiver incremental tuning. It is your receiver's "fine tuning" function. If the voice your receiving sounds a bit too soprano or a bit too bass than what is natural, try turning the RIT until the voice sounds more human.

T4B07 What does the term "RIT" mean?
Answer: *Receiver Incremental Tuning.*

RIT is the receiver's "fine tuning" circuit.

T4B08 What is the advantage of having multiple receive bandwidth choices on a multimode transceiver?
Answer: *Permits noise or interference reduction by selecting a bandwidth matching the mode.*

These bandwidth choices are filters whose bandwidths match the more popular modes. There is usually one labeled for SSB, CW, FM, AM, etc. For instance, a filter with a 2400 Hz bandwidth helps eliminate unwanted noise when listening for SSB signals. For CW it is 500 Hz.

T4B09 Which of the following is an appropriate receive filter to select in order to minimize noise and interference for SSB reception?
Answer: *2400 Hz.*

A 2400 Hz filter will minimize excess noise and interferencc in SSB reception.

T4B10 Which of the following is an appropriate receive filter to select in order to minimize noise and interference for CW reception?
Answer: *500 Hz.*

Same idea as the previous question. A 500 Hz filter will focus your reception just on CW signals.

T4B11 Which of the following describes the common meaning of the term "repeater offset"?
Answer: *The difference between the repeater's transmit and receive frequencies.*

The offset of a repeater is the difference between the repeater's receiving frequency and transmitting frequency. The offset is usually described as "plus or minus" as the offset for the transmit frequency may be either higher or lower than the receive frequency. For instance, the common offset for the 2 meter band is plus or minus 600 kHz.

Quiz #14

(Find the answers on page 256)

1. _____ T4B01 What may happen if a transmitter is operated with the microphone gain set too high?
A. The output power might be too high
B. The output signal might become distorted
C. The frequency might vary
D. The SWR might increase

2. _____ T4B02 Which of the following can be used to enter the operating frequency on a modern transceiver?
A. The keypad or VFO knob
B. The CTCSS or DTMF encoder
C. The Automatic Frequency Control
D. All of these choices are correct

3. _____ T4B03 What is the purpose of the squelch control on a transceiver?
A. To set the highest level of volume desired
B. To set the transmitter power level
C. To adjust the automatic gain control
D. To mute receiver output noise when no signal is being received

4. _____ T4B04 What is a way to enable quick access to a favorite frequency on your transceiver?
A. Enable the CTCSS tones
B. Store the frequency in a memory channel
C. Disable the CTCSS tones
D. Use the scan mode to select the desired frequency

5. _____ T4B05 Which of the following would reduce ignition interference to a receiver?
A. Change frequency slightly
B. Decrease the squelch setting
C. Turn on the noise blanker
D. Use the RIT control

6. _____ T4B06 Which of the following controls could be used if the voice pitch of a single-sideband signal seems too high or low?
A. The AGC or limiter
B. The bandwidth selection
C. The tone squelch
D. The receiver RIT or clarifier

7. _____ T4B07 What does the term "RIT" mean?
A. Receiver Input Tone
B. Receiver Incremental Tuning
C. Rectifier Inverter Test
D. Remote Input Transmitter

8. _____ T4B08 What is the advantage of having multiple receive bandwidth choices on a multimode transceiver?
A. Permits monitoring several modes at once
B. Permits noise or interference reduction by selecting a bandwidth matching the mode
C. Increases the number of frequencies that can be stored in memory
D. Increases the amount of offset between receive and transmit frequencies

9. _____ T4B09 Which of the following is an appropriate receive filter to select in order to minimize noise and interference for SSB reception?
A. 500 Hz
B. 1000 Hz
C. 2400 Hz
D. 5000 Hz

10. _____ T4B10 Which of the following is an appropriate receive filter to select in order to minimize noise and interference for CW reception?
A. 500 Hz
B. 1000 Hz
C. 2400 Hz
D. 5000 Hz

11. _____ T4B11 Which of the following describes the common meaning of the term “repeater offset”?
A. The distance between the repeater’s transmit and receive antennas
B. The time delay before the repeater timer resets
C. The difference between the repeater’s transmit and receive frequencies
D. The maximum frequency deviation permitted on the repeater’s input signal

Lesson 14 Study Sheet

✓ RIT stands for receiver incremental tuning. It is the receiver's fine tuning circuit.

✓ An appropriate receive filter for SSB is 2400 Hz.

✓ An appropriate receive filter for CW is 500 Hz.

Notes:

Lesson 15
The T5A Section
Electrical Principles

T5A01 Electrical current is measured in which of the following units?
Answer: *Amperes.*

Amperes (A) is the unit of measure for electrical current. Electrical current is the flow of electrons in an electric circuit.

T5A02 Electrical power is measured in which of the following units?
Answer: *Watts.*

Watts (W) is the unit of measure of power. Power is the rate at which energy is used.

T5A03 What is the name for the flow of electrons in an electric circuit?
Answer: *Current.*

Current is the flow of electrons in an electric circuit. Current is measured in amperes.

T5A04 What is the name for a current that flows only in one direction?
Answer: *Direct current.*

Direct current, or DC, is current that only flows in one direction. A battery is a common source of direct current.

T5A05 What is the electrical term for the electromotive force (EMF) that causes electron flow?
Answer: *Voltage.*

Voltage (V) is also known as electromotive force (EMF). If you think of electric current flowing through a wire as water flowing through a pipe, voltage is the pressure pushing the water through the pipe. Voltage is the force, or electromotive force to be exact, that pushes electrons through a wire or circuit.

T5A06 How much voltage does a mobile transceiver usually require?
Answer: *About 12 volts.*

Where do you usually use a mobile radio? In a vehicle. What is the common voltage of a car battery? About 12 Volts.

T5A07 Which of the following is a good electrical conductor?
Answer: *Copper.*

Conductors are materials that allow electric current to flow freely through them. Metals are good conductors. The atoms in good conductors share their electrons relatively easily with other atoms. Voltage is the force that causes these atoms to give up there electrons and pushes the electric current. Materials which gives up their electrons with a very little amount of voltage are good conductors.

T5A08 Which of the following is a good electrical insulator?
Answer: *Glass.*

Insulators do not conduct electricity well. Materials which are good insulators do not give up their electrons very easily. It takes an extremely high voltage for an insulator to allow electrical current to flow through them. Glass, air, wood, and rubber are good insulators.

T5A09 What is the name for a current that reverses direction on a regular basis?
Answer: *Alternating current.*

Alternating current, or AC, reverses its direction of current on a regular basis. Current will flow in one direction and then switch directions. The other alternative is direct current, or DC, which is current that flows only in one direction.

T5A10 Which term describes the rate at which electrical energy is used?
Answer: *Power.*

Power is the rate at which energy is used.

T5A11 What is the basic unit of electromotive force?
Answer: *The volt.*

Electromotive force (EMF) is also known as voltage. The basic unit of voltage is the volt.

Quiz #15

(Find the answers on page 256)

1. _____ T5A01 Electrical current is measured in which of the following units?
A. Volts
B. Watts
C. Ohms
D. Amperes

2. _____ T5A02 Electrical power is measured in which of the following units?
A. Volts
B. Watts
C. Ohms
D. Amperes

3. _____ T5A03 What is the name for the flow of electrons in an electric circuit?
A. Voltage
B. Resistance
C. Capacitance
D. Current

4. _____ T5A04 What is the name for a current that flows only in one direction?
A. Alternating current
B. Direct current
C. Normal current
D. Smooth current

5. _____ T5A05 What is the electrical term for the electromotive force (EMF) that causes electron flow?
A. Voltage
B. Ampere-hours
C. Capacitance
D. Inductance

6. _____ T5A06 How much voltage does a mobile transceiver usually require?
A. About 12 volts
B. About 30 volts
C. About 120 volts
D. About 240 volts

7. _____ T5A07 Which of the following is a good electrical conductor?
A. Glass
B. Wood
C. Copper
D. Rubber

8. _____ T5A08 Which of the following is a good electrical insulator?
A. Copper
B. Glass
C. Aluminum
D. Mercury

9. _____ T5A09 What is the name for a current that reverses direction on a regular basis?
A. Alternating current
B. Direct current
C. Circular current
D. Vertical current

10. _____ T5A10 Which term describes the rate at which electrical energy is used?
A. Resistance
B. Current
C. Power
D. Voltage

11. _____ T5A11 What is the basic unit of electromotive force?
A. The volt
B. The watt
C. The ampere
D. The ohm

Chapter 15 Study Sheet

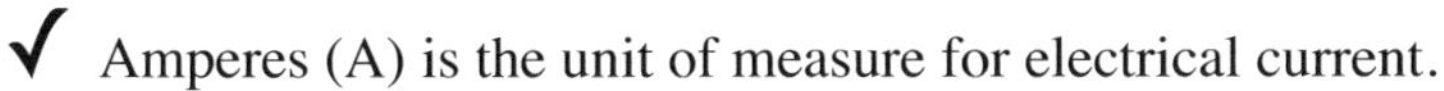

✓ Amperes (A) is the unit of measure for electrical current.

✓ Watts (W) is the unit of measure of power.

✓ Volts (V) are the basic unit of electromotive force, or voltage.

✓ Current is the flow of electrons in an electric circuit.

✓ Direct current, or DC, is current that only flows in one direction.

✓ Alternating current, or AC, reverses its direction of current on a regular basis.

Notes:

Lesson 16
The T5B Section
Math for Electronics
(You will need a scientific calculator.)

T5B01 How many milliamperes is 1.5 amperes?
Answer: *1,500 milliamperes.*

The metric prefix "milli" means 1 / 1000th of something. For this question, that something is amperes. To find milliamperes, just multiply 1.5 amperes by the number of milliamperes in an ampere. 1.5 amperes × 1000 = 1500 milliamperes.

T5B02 What is another way to specify a radio signal frequency of 1,500,000 hertz?
Answer: *1500 kHz.*

This question is a little trickier than the last. For this question you need to know the value of the metric prefixes "kilo," "mega," and "giga" to figure the answer for all the possible answers on the exam. "Kilo" is 1000 of something. "Mega" is 1,000,000 of something. "Giga" is 1,000,000,000 of something. To go from smaller units to larger units in the metric system, you divide. For instance, to find megahertz, divide the number of hertz by 1,000,000. 1,500,000 Hz / 1,000,000 = 1.5 MHz. To figure how many kilohertz (kHz) are in 1,500,000 hertz, you need to divide by 1000 (1 "kilo"). 1,500,000 hertz / 1000 is 1500 kHz, the answer.

T5B03 How many volts are equal to one kilovolt?
Answer: *One thousand volts.*

You already know that "kilo" is 1000 of something. One kilovolt is 1000 volts!

T5B04 How many volts are equal to one microvolt?
Answer: *One one-millionth of a volt.*

The metric prefix "micro" is 1 / 1,000,000th of something. One microvolt equals 1 / 1,000,000th of a volt which is also expressed as 0.000001 volts.

T5B05 Which of the following is equivalent to 500 milliwatts?
Answer: *0.5 watts.*

You already know that the metric prefix "milli" means 1 / 1000th of something. Another way of expressing it is a "milli" is .001. To find watts, you can either divide 500 milliwatts by 1000 or you can multiply 500 milliwatts by .001 (which is 1 divided by 1000, 1 ÷ 1000 = .001). Either way, you get 0.5 watts. You can use whatever method you are most comfortable with.

T5B06 If an ammeter calibrated in amperes is used to measure a 3000-milliampere current, what reading would it show?
Answer: *3 amperes.*

For the purposes of this question, an ammeter measures current in amperes. Again we are going from a small unit of measure to a larger, milliamperes to amperes. This means we can divide 3000 milliamperes by 1000 (1000 milliamperes per ampere), or you can multiply 3000 milliamperes by .001. Which ever way you choose, you get 3 amperes.

T5B07 If a frequency readout calibrated in megahertz shows a reading of 3.525 MHz, what would it show if it were calibrated in kilohertz?
Answer: *3525 kHz.*

To go from MHz (megahertz) to kHz (kilohertz) you are going from a larger unit of measure to a smaller unit of measure. You know that there are a 1,000,000 hertz in one megahertz and there are 1000 hertz in one kilohertz. One way to solve this problem is to convert 3.525 MHz into Hz and then calculate for kHz. To solve the problem this way, multiply 3.525 MHz by 1,000,000 (there are 1,000,000 Hz in a MHz) and get 3,525,000 Hz. To get kHz you divide 3,525,000 Hz by 1000 (there are 1000 Hz in a kHz) which equals 3525 kHz. A faster way to solve this problem is figure how many kHz are in 1 MHz and multiply that number by 3.525 MHz. There are 1000 kHz in 1 MHz. Multiply 1000 by 3.525 MHz and you will get 3525 kHz.

T5B08 How many microfarads are 1,000,000 picofarads?
Answer: *1 microfarad.*

The metric prefix "pico" is one-trillionth of something. A farad is the basic unit of capacitance. A microfarad is one-millionth of a farad. Let's solve for farads first. You are going from a smaller unit of measure to a larger so you will divide. Divide 1,000,000 by one trillion and you get 0.000001 farad. Now to solve for microfarads you are going from a larger unit of measure (farad) to a smaller (microfarad) so you will multiply. Multiply 0.000001 farads by 1,000,000 and you get 1 microfarad. You have to punch a lot of zeros into your calculator to use this method. If you understand powers of ten, this may be a lot easier. A picofarad can be expressed as 1×10^{-12} farads. A microfarad can also be expressed as 1×10^{-6} farads. To express 1,000,000 picofarads this way you first need to express 1,000,000 as a power of ten. 1,000,000 is equal to 1×10^{6}. Take the number of picofarads (1×10^{6}) and multiply by the number of picofarads in a farad (1×10^{-12}). You get 1×10^{-6} farads or 1 microfarad. Be sure you check the instruction manual for your scientific calculator for specific instructions on how to calculate powers of ten.

T5B09 What is the approximate amount of change, measured in decibels (dB), of a power increase from 5 watts to 10 watts?
Answer: *3 dB.*

Decibels, abbreviated dB, is a measurement of change from a reference power, or intensity, to a new power, or intensity. For instance, if you are listening to the radio at a certain volume and decide to turn the volume up, there has been a change in volume that can be measured in dB. There is a formula that you will need to know. That formula is:

$$dB = 10 \log (P_1/P_0)$$

P_0 is the reference power. This is the power we start with in the above question which is 5 watts. P_1 is the power level compared to the reference which is the power we end with in the above question, 10 watts. Now let's plug the numbers into the formula.

$$dB = 10 \log (P_1/P_0)$$
$$dB = 10 \log (10 \text{ watts} / 5 \text{ watts})$$
$$dB = 10 \log (2)$$
$$dB = 10 (0.301)$$
$$dB = 3.01 \text{ or approximately } 3 \text{ dB}$$

T5B10 What is the approximate amount of change, measured in decibels (dB), of a power decrease from 12 watts to 3 watts?
Answer: *6 dB.*

Let's use the decibel formula.

$$dB = 10 \log (P_1/P_0)$$
$$dB = 10 \log (3 \text{ watts} / 12 \text{ watts})$$
$$dB = 10 \log (0.25)$$
$$dB = 10 (-0.602)$$
$$dB = -6.02$$

A negative dB just means the new power is less than the reference power it is compared to. The same way you cannot travel a negative distance, change in decibels is usually not expressed as negative either. Therefore the answer is approximately 6 dB.

T5B11 What is the approximate amount of change, measured in decibels (dB), of a power increase from 20 watts to 200 watts?
Answer: *10 dB*.

Let's use the formula one more time.

$$dB = 10 \log (P_1/P_0)$$
$$dB = 10 \log (200 \text{ watts} / 20 \text{ watts})$$
$$dB = 10 \log (10)$$
$$dB = 10 (1)$$
$$dB = 10 \text{ or } 10 \text{ dB}$$

Quiz #16

(Find the answers on page 256)

1. _____ T5B01 How many milliamperes is 1.5 amperes?
A. 15 milliamperes
B. 150 milliamperes
C. 1,500 milliamperes
D. 15,000 milliamperes

2. _____ T5B02 What is another way to specify a radio signal frequency of 1,500,000 hertz?
A. 1500 kHz
B. 1500 MHz
C. 15 GHz
D. 150 kHz

3. _____ T5B03 How many volts are equal to one kilovolt?
A. One one-thousandth of a volt
B. One hundred volts
C. One thousand volts
D. One million volts

4. _____ T5B04 How many volts are equal to one microvolt?
A. One one-millionth of a volt
B. One million volts
C. One thousand kilovolts
D. One one-thousandth of a volt

5. _____ T5B05 Which of the following is equivalent to 500 milliwatts?
A. 0.02 watts
B. 0.5 watts
C. 5 watts
D. 50 watts

6. _____ T5B06 If an ammeter calibrated in amperes is used to measure a 3000-milliampere current, what reading would it show?
A. 0.003 amperes
B. 0.3 amperes
C. 3 amperes
D. 3,000,000 amperes

7. _____ T5B07 If a frequency readout calibrated in megahertz shows a reading of 3.525 MHz, what would it show if it were calibrated in kilohertz?
A. 0.003525 kHz
B. 35.25 kHz
C. 3525 kHz
D. 3,525,000 kHz

8. _____ T5B08 How many microfarads are 1,000,000 picofarads?
A. 0.001 microfarads
B. 1 microfarad
C. 1000 microfarads
D. 1,000,000,000 microfarads

9. _____ T5B09 What is the approximate amount of change, measured in decibels (dB), of a power increase from 5 watts to 10 watts?
A. 2 dB
B. 3 dB
C. 5 dB
D. 10 dB

10. _____ T5B10 What is the approximate amount of change, measured in decibels (dB), of a power decrease from 12 watts to 3 watts?
A. 1 dB
B. 3 dB
C. 6 dB
D. 9 dB

11. _____ T5B11 What is the approximate amount of change, measured in decibels (dB), of a power increase from 20 watts to 200 watts?
A. 10 dB
B. 12 dB
C. 18 dB
D. 28 dB

Lesson 16 Study Sheet

✓ The metric prefixes you will need to know for the exam:

giga = 1×10^9 = 1,000,000,000
mega = 1×10^6 = 1,000,000
kilo = 1×10^3 = 1000
milli = 1×10^{-3} = 0.001
micro = 1×10^{-6} = 0.000001
pico = 1×10^{-12} = 0.000000000001

✓ The decibel (dB) formula:

$$dB = 10 \log (P_1/P_0)$$

P_0 is the reference power and P_1 is the power level compared to the reference.

Notes:

Lesson 17
The T5C Section
Electronic Principles

T5C01 What is the ability to store energy in an electric field called?
Answer: *Capacitance.*

Capacitance is the ability to store energy in an electric field. In electronics, this is done with a component called a capacitor. A capacitor consists of two plates separated by a very small amount of space, but not touching. One plate is connected to the positive end of a battery or power source and the other plate is connected to the negative. When the power is turned on, the plate attached to the positive connection becomes positively charged and the plate attached to the negative end becomes negatively charged. These two oppositely charged plates create an electric field between them which is used to store energy.

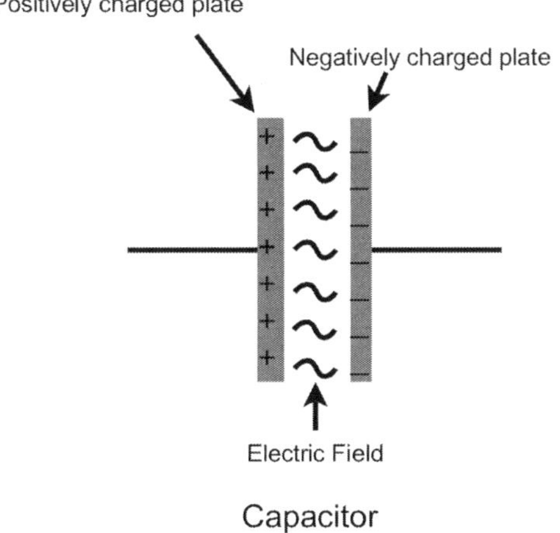

Capacitor

T5C02 What is the basic unit of capacitance?
Answer: *The farad.*

The farad is the basic unit of capacitance.

T5C03 What is the ability to store energy in a magnetic field called?
Answer: *Inductance.*

Inductance is the ability to store energy in a magnetic field. The electrical component is called an inductor. An inductor is basically a coil of wire with an electric current running through it. An increase in current through an inductor produces a magnetic field around the inductor. This magnetic field stores energy.

T5C04 What is the basic unit of inductance?
Answer: *The henry.*

The henry is the basic unit of inductance.

T5C05 What is the unit of frequency?
Answer: *Hertz.*

This one should be easy by now! Hertz is the basic unit of frequency.

T5C06 What is the abbreviation that refers to radio frequency signals of all types?
Answer: *RF.*

RF stands for radio frequency. For instance, RF signals are radio frequency signals. RF radiation is radio frequency radiation. RFI is radio frequency interference.

T5C07 What is a usual name for electromagnetic waves that travel through space?
Answer: *Radio waves.*

A few lessons ago we talked about how radio waves are electromagnetic waves composed of electric and magnetic fields.

T5C08 What is the formula used to calculate electrical power in a DC circuit?
Answer: *Power (P) equals voltage (E) multiplied by current (I).*

The formula used to calculate power in a DC circuit is P=IE...PIE! P is power expressed in watts, I is current expressed in amperes, and E is voltage expressed in volts.

T5C09 How much power is being used in a circuit when the applied voltage is 13.8 volts DC and the current is 10 amperes?
Answer: *138 watts.*

Use the power equation P=IE to solve this one.

P (in watts) = I (in amperes) x E (in volts)
P = IE
P = (10 amperes)(13.8 volts)
P = 138 watts

T5C10 How much power is being used in a circuit when the applied voltage is 12 volts DC and the current is 2.5 amperes?
Answer: *30 watts.*

Here is the power equation again!

P (in watts) = I (in amperes) x E (in volts)
P = IE
P = (2.5 amperes)(12 volts)
P = 30 watts

T5C11 How many amperes are flowing in a circuit when the applied voltage is 12 volts DC and the load is 120 watts?
Answer: *10 amperes.*

You can change the P=IE equation around to solve for any of the variables. For this question you need to solve for amperes or "I" in the equation. You can solve for "I" by dividing both sides of the equation by "E."

P = IE
P/E = IE/E
P/E = I
I = P/E

A quick way to solve for any of the variables in the P=IE equation is to use the P=IE circle. Cover the variable you want to find with your finger. The position of the remaining letters gives the correct equation. For this question, place your finger over the "I" in the circle. What is left is "P/E." If you want to solve for "E," place your finger over the "E" and what is left is "P/I."

Let's answer the question with the equation which solves for amperes (I).

I (in amperes) = P (in watts)/E (in volts)
I = P/E
I = (120 watts)/(12 volts)
I = 10 amperes

Quiz #17

(Find the answers on page 256)

1. _____ T5C01 What is the ability to store energy in an electric field called?
A. Inductance
B. Resistance
C. Tolerance
D. Capacitance

2. _____ T5C02 What is the basic unit of capacitance?
A. The farad
B. The ohm
C. The volt
D. The henry

3. _____ T5C03 What is the ability to store energy in a magnetic field called?
A. Admittance
B. Capacitance
C. Resistance
D. Inductance

4. _____ T5C04 What is the basic unit of inductance?
A. The coulomb
B. The farad
C. The henry
D. The ohm

5. _____ T5C05 What is the unit of frequency?
A. Hertz
B. Henry
C. Farad
D. Tesla

6. _____ T5C06 What is the abbreviation that refers to radio frequency signals of all types?
A. AF
B. HF
C. RF
D. VHF

7. _____ T5C07 What is a usual name for electromagnetic waves that travel through space?
A. Gravity waves
B. Sound waves
C. Radio waves
D. Pressure waves

8. _____ T5C08 What is the formula used to calculate electrical power in a DC circuit?
A. Power (P) equals voltage (E) multiplied by current (I)
B. Power (P) equals voltage (E) divided by current (I)
C. Power (P) equals voltage (E) minus current (I)
D. Power (P) equals voltage (E) plus current (I)

9. _____ T5C09 How much power is being used in a circuit when the applied voltage is 13.8 volts DC and the current is 10 amperes?
A. 138 watts
B. 0.7 watts
C. 23.8 watts
D. 3.8 watts

10. _____ T5C10 How much power is being used in a circuit when the applied voltage is 12 volts DC and the current is 2.5 amperes?
A. 4.8 watts
B. 30 watts
C. 14.5 watts
D. 0.208 watts

11. _____ T5C11 How many amperes are flowing in a circuit when the applied voltage is 12 volts DC and the load is 120 watts?
A. 0.1 amperes
B. 10 amperes
C. 12 amperes
D. 132 amperes

Lesson 17 Study Sheet

✓ Capacitance is the ability to store energy in an electric field.

✓ Inductance is the ability to store energy in a magnetic field.

✓ The farad is the basic unit of capacitance.

✓ The henry is the basic unit of inductance.

✓ Hertz is the basic unit of frequency.

✓ P=IE is the question used to calculate electric power in a DC Circuit.

Notes:

Lesson 18
The T5D Section
Ohm's Law

Ohm's Law describes the relationship between voltage, current, and resistance in a circuit. Ohm's Law states that voltage is equal to the current multiplied by the resistance. The basic equation is:

E (in volts) = I (in amperes) x R (in ohms)

We have talked a bit about voltage and current in previous lessons, but not resistance. Resistance is the opposition to electric current. All material exhibits resistance to some extent, even the best conductors. Put to the right use, this resistance can come in very handy in electronics. Resistors are an electronic component that are made with the intent of providing resistance in a circuit. The basic unit of measure for resistance is the ohm.

Like the equation to calculate electric power in a DC circuit, P=IE, you can use the Ohm's Law circle to solve for any of the variables in Ohm's Law.

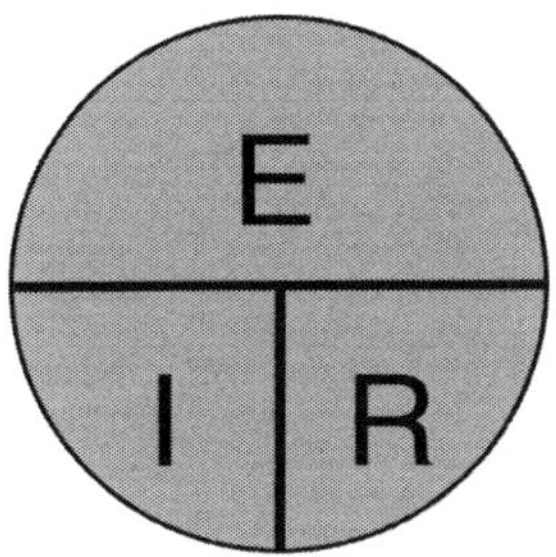

T5D01 What formula is used to calculate current in a circuit?
Answer: *Current (I) equals voltage (E) divided by resistance (R).*

The first few questions in this lesson deal with the various forms of Ohm's Law, solving for voltage, current, or resistance. You can either use the Ohm's Law circle or solve it the long way. This question is looking for current (I). Let's start with the basic equation and solve for current (I).

E = IR Now divide both sides by "R"
E/R = IR/R
E/R = I
I = E/R
Current (I) equals Voltage (E) divided by Resistance (R)

T5D02 What formula is used to calculate voltage in a circuit?
Answer: *Voltage (E) equals current (I) multiplied by resistance (R).*

This question is looking for voltage (E). The answer is the basic equation.

E = IR
Voltage (E) equals Current (I) multiplied by Resistance (R)

T5D03 What formula is used to calculate resistance in a circuit?
Answer: *Resistance (R) equals voltage (E) divided by current (I).*

This question is looking for resistance (R).

E = IR Now divide both sides by "I"
E/I = IR/I
E/I = R
R = E/I
Resistance (R) equals Voltage (E) divided by Current (I)

T5D04 What is the resistance of a circuit in which a current of 3 amperes flows through a resistor connected to 90 volts?
Answer: *30 ohms.*

Let's put Ohm's Law into practice. This question is looking for the resistance in the circuit. You can use the version of Ohm's Law you solved for resistance (R) above.

R (in ohms) = E (in volts)/I (in amperes)
R = E/I
R = (90 volts)/(3 amperes)
R = 30 ohms

T5D05 What is the resistance in a circuit for which the applied voltage is 12 volts and the current flow is 1.5 amperes?
Answer: *8 ohms.*

Use the same version of Ohm's Law solving for resistance (R).

R (in ohms) = E (in volts)/I (in amperes)
R = E/I
R = (12 volts)/(1.5 amperes)
R = 8 ohms

T5D06 What is the resistance of a circuit that draws 4 amperes from a 12-volt source?
Answer: *3 ohms.*

Once again, use the same version of Ohm's Law solving for resistance (R).

R (in ohms) = E (in volts)/I (in amperes)
R = E/I
R = (12 volts)/(4 amperes)
R = 3 ohms

T5D07 What is the current flow in a circuit with an applied voltage of 120 volts and a resistance of 80 ohms?
Answer: *1.5 amperes.*

This question is looking for current (I). Use the version of Ohm's Law solving for current (I) we worked out in the first question.

I (amperes) = E (volts)/R (ohms)
I = E/R
I = (120 volts)/(80 ohms)
I = 1.5 amperes

T5D08 What is the current flowing through a 100-ohm resistor connected across 200 volts?
Answer: *2 amperes.*

Here is the current (I) version of Ohm's Law again.

I (amperes) = E (volts)/R (ohms)
I = E/R
I = (200 volts)/(100 ohms)
I = 2 amperes

T5D09 What is the current flowing through a 24-ohm resistor connected across 240 volts?
Answer: *10 amperes.*

One more time solving for current.

I (amperes) = E (volts)/R (ohms)
I = E/R
I = (240 volts)/(24 ohms)
I = 10 amperes

T5D10 What is the voltage across a 2-ohm resistor if a current of 0.5 amperes flows through it?
Answer: *1 volt.*

This question is looking for the voltage (E) of the circuit. You can use the basic Ohm's Law equation to solve for voltage (E).

E (in volts) = I (in amperes) x R (in ohms)
E = IR
E = (0.5 amperes) x (2 ohm)
E = 1 volt

T5D11 What is the voltage across a 10-ohm resistor if a current of 1 ampere flows through it?
Answer: *10 volts.*

E (in volts) = I (in amperes) x R (in ohms)
E = IR
E = (1 ampere) x (10 ohms)
E = 10 volts

T5D12 What is the voltage across a 10-ohm resistor if a current of 2 amperes flows through it?
Answer: 20 volts.

E (in volts) = I (in amperes) x R (in ohms)
E = IR
E = (2 amperes) x (10 ohms)
E = 20 volts

Quiz #18

(Find the answers on page 256)

1. _____ T5D01 What formula is used to calculate current in a circuit?
A. Current (I) equals voltage (E) multiplied by resistance (R)
B. Current (I) equals voltage (E) divided by resistance (R)
C. Current (I) equals voltage (E) added to resistance (R)
D. Current (I) equals voltage (E) minus resistance (R)

2. _____ T5D02 What formula is used to calculate voltage in a circuit?
A. Voltage (E) equals current (I) multiplied by resistance (R)
B. Voltage (E) equals current (I) divided by resistance (R)
C. Voltage (E) equals current (I) added to resistance (R)
D. Voltage (E) equals current (I) minus resistance (R)

3. _____ T5D03 What formula is used to calculate resistance in a circuit?
A. Resistance (R) equals voltage (E) multiplied by current (I)
B. Resistance (R) equals voltage (E) divided by current (I)
C. Resistance (R) equals voltage (E) added to current (I)
D. Resistance (R) equals voltage (E) minus current (I)

4. _____ T5D04 What is the resistance of a circuit in which a current of 3 amperes flows through a resistor connected to 90 volts?
A. 3 ohms
B. 30 ohms
C. 93 ohms
D. 270 ohms

5. _____ T5D05 What is the resistance in a circuit for which the applied voltage is 12 volts and the current flow is 1.5 amperes?
A. 18 ohms
B. 0.125 ohms
C. 8 ohms
D. 13.5 ohms

6. _____ T5D06 What is the resistance of a circuit that draws 4 amperes from a 12-volt source?
A. 3 ohms
B. 16 ohms
C. 48 ohms
D. 8 Ohms

7. _____ T5D07 What is the current flow in a circuit with an applied voltage of 120 volts and a resistance of 80 ohms?
A. 9600 amperes
B. 200 amperes
C. 0.667 amperes
D. 1.5 amperes

8. _____ T5D08 What is the current flowing through a 100-ohm resistor connected across 200 volts?
A. 20,000 amperes
B. 0.5 amperes
C. 2 amperes
D. 100 amperes

9. _____ T5D09 What is the current flowing through a 24-ohm resistor connected across 240 volts?
A. 24,000 amperes
B. 0.1 amperes
C. 10 amperes
D. 216 amperes

10. _____ T5D10 What is the voltage across a 2-ohm resistor if a current of 0.5 amperes flows through it?
A. 1 volt
B. 0.25 volts
C. 2.5 volts
D. 1.5 volts

11. _____ T5D11 What is the voltage across a 10-ohm resistor if a current of 1 ampere flows through it?
A. 1 volt
B. 10 volts
C. 11 volts
D. 9 volts

12. _____ T5D12 What is the voltage across a 10-ohm resistor if a current of 2 amperes flows through it?
A. 8 volts
B. 0.2 volts
C. 12 volts
D. 20 volts

Lesson 18 Study Sheet

✓ Ohm's Law states voltage (E) equals current (I) multiplied by the resistance (R).

E (in volts) = I (in amperes) x R (in ohms)

E = IR

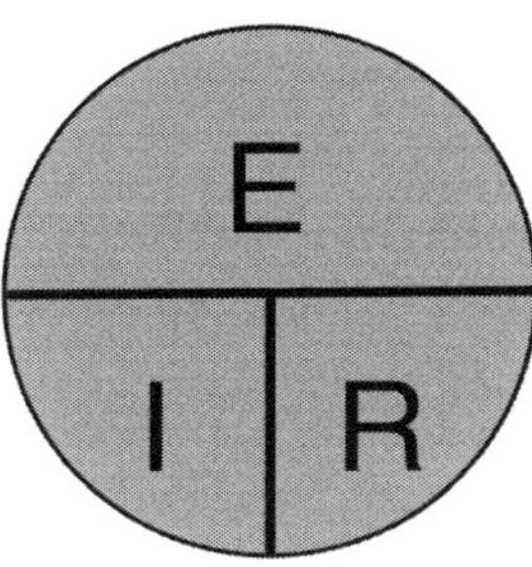

Notes:

Lesson 19
The T6A Section
Electrical Components

T6A01 What electrical component is used to oppose the flow of current in a DC circuit?
Answer: *Resistor.*

A resistor is an electrical component that opposes, or resists, the flow of current in a circuit. Resistors have a measured resistance and vary in size and shape. The name says it all. Resistors oppose, or resist, the flow of current.

T6A02 What type of component is often used as an adjustable volume control?
Answer: *Potentiometer.*

A potentiometer is a type of resistor where the level of resistance can be adjusted. Potentiometers are often used as volume controls. As you turn the knob you can either decrease the resistance which increases the current flow resulting in an increase in volume, or increase the resistance which reduces the current flow and lowers the volume.

T6A03 What electrical parameter is controlled by a potentiometer?
Answer: *Resistance.*

A potentiometer is used to adjust resistance.

T6A04 What electrical component stores energy in an electric field?
Answer: *Capacitor.*

Capacitance is the ability to store energy in an electric field. The electrical component that handles capacitance is a capacitor. A capacitor is two or more metal plates separated by a very small amount of space. The space between the two plates is often filled by an insulating material (air can be considered a type of insulator as well). The number of plates, size of the plates, and the type of insulating material separating the plates all have an impact on the level of capacitance.

T6A05 What type of electrical component consists of two or more conductive surfaces separated by an insulator?
Answer: *Capacitor.*

Here is the capacitor again. A capacitor consists of two or more conductive surfaces, or plates, separated by an insulating material.

T6A06 What type of electrical component stores energy in a magnetic field?
Answer: *Inductor.*

Inductors are generally coils of wire that produce a magnetic field when electric current flows through them. Energy is stored in an inductor's magnetic field. Sometimes inductors are wrapped around a core of conductive material. The number of turns in the coil, the presence or absence of a core, and the material the core is made of all help determine the level of inductance for each particular inductor.

T6A07 What electrical component is usually composed of a coil of wire?
Answer: *Inductor.*

Inductors consist of coils of wire. These coils come in many shapes and sizes.

T6A08 What electrical component is used to connect or disconnect electrical circuits?
Answer: *Switch.*

Don't over think the question! A switch turns things on and off by connecting or disconnecting the flow of electricity through a circuit.

T6A09 What electrical component is used to protect other circuit components from current overloads?
Answer: *Fuse.*

Fuses are in cars, houses, and a lot of electrical equipment. A fuse is a disposable component designed to protect electronic circuits from damage from too much current. Most fuses have a strip of conductive material that is designed to burn out when the current exceeds a certain level. When the fuse burns out it stops the flow of electricity through the circuit.

T6A10 What is the nominal voltage of a fully charged nickel-cadmium cell?
Answer: *1.2 volts*.

Unless you are already familiar with nickel-cadmium, or NiCad, batteries you should probably memorize the answer to this question. The normal voltage of a charged nickel-cadmium battery is 1.2 volts.

T6A11 Which battery type is not rechargeable?
Answer: *Carbon-zinc*.

These are regular disposable batteries. This answer is another one you should probably memorize. You may also be able to find the correct answer by eliminating the other possible answers. The other possible answers are: nickel-cadmium, lead-acid, and lithium-ion. Nickel-cadmium, or NiCad, batteries are found in many radio controlled toys and small electronics that can be recharged. Lead-acid batteries are commonly used as car batteries which are recharged every time the engine is runs. Lithium-ion batteries are commonly used in laptop computers and cell phones and, again, are rechargeable. Of the possible answers, Carbon-zinc batteries are disposable and cannot be recharged.

Quiz #19

(Find the answers on page 256)

1. _____ T6A01 What electrical component is used to oppose the flow of current in a DC circuit?
A. Inductor
B. Resistor
C. Voltmeter
D. Transformer

2. _____ T6A02 What type of component is often used as an adjustable volume control?
A. Fixed resistor
B. Power resistor
C. Potentiometer
D. Transformer

3. _____ T6A03 What electrical parameter is controlled by a potentiometer?
A. Inductance
B. Resistance
C. Capacitance
D. Field strength

4. _____ T6A04 What electrical component stores energy in an electric field?
A. Resistor
B. Capacitor
C. Inductor
D. Diode

5. _____ T6A05 What type of electrical component consists of two or more conductive surfaces separated by an insulator?
A. Resistor
B. Potentiometer
C. Oscillator
D. Capacitor

6. _____ T6A06 What type of electrical component stores energy in a magnetic field?
A. Resistor
B. Capacitor
C. Inductor
D. Diode

7. _____ T6A07 What electrical component is usually composed of a coil of wire?
A. Switch
B. Capacitor
C. Diode
D. Inductor

8. _____ T6A08 What electrical component is used to connect or disconnect electrical circuits?
A. Zener Diode
B. Switch
C. Inductor
D. Variable resistor

9. _____ T6A09 What electrical component is used to protect other circuit components from current overloads?
A. Fuse
B. Capacitor
C. Shield
D. Inductor

10. _____ T6A10 What is the nominal voltage of a fully charged nickel-cadmium cell?
A. 1.0 volts
B. 1.2 volts
C. 1.5 volts
D. 2.2 volts

11. _____ T6A11 Which battery type is not rechargeable?
A. Nickel-cadmium
B. Carbon-zinc
C. Lead-acid
D. Lithium-ion

Lesson 19 Study Sheet

✓ A resistor is an electrical component that opposes, or resists, the flow of current in a circuit.

✓ A potentiometer is a type of component often used as an adjustable volume control. A potentiometer adjusts resistance in a circuit.

✓ A capacitor is a type of component that consists of two or more conductive surfaces separated by an insulator and stores energy in an electric field.

✓ An inductor is a type of component that stores energy in a magnetic field and is usually composed of a coil of wire.

✓ The nominal voltage of a fully charged nickel-cadmium cell is 1.2 volts.

✓ Carbon-zinc batteries are disposable and cannot be recharged.

Notes:

Lesson 20
The T6B Section
Semiconductors

Semiconductors are electronic components that are made of semiconductor material. This semiconductor material is neither a good conductor or a good insulator, it is somewhere in the middle. The two most commonly used materials in semiconductor components are silicon crystals and germanium crystals. A semiconductor manufacturer will treat, or dope, these crystals with a chemical. Depending on the type of chemical used, the resulting doped material will either have a positive bias or a negative bias. The positively biased material is called P type material. The negatively biased material is called N type material. There are several different ways these P type and N type materials are configured with each other. When current is run through these various configurations, they produce some very valuable electronic functions. Depending on the manner which the P type and N type material are combined, semiconductors can act as switches, amplifiers, or limit electron flow to only one direction.

T6B01 What class of electronic components is capable of using a voltage or current signal to control current flow?
Answer: *Transistors.*

Transistors can control current flow by using a voltage or current signal. To describe how this works we will use a bipolar transistor as an example. Transistors consist of three layers of semiconductor material. For this example, the three layers consist of one N type layer of material sandwiched between two P type material layers.

Transistor

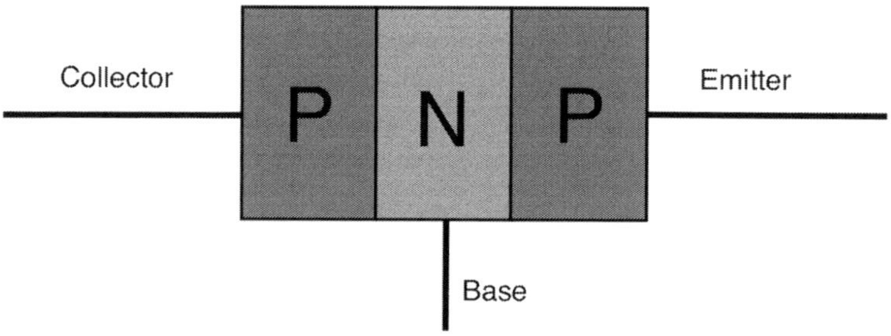

Each of the three layers of material has its own electrode. The names of the electrodes are the emitter, collector, and base. In this example, the emitter is connected to the positive end of the power source and feeds into one of the P type materials on the end. The collector will be connected to the negative end of the power source and is connected to the P type material opposite the emitter. The base is connected to a negative power source and feeds into the N type material in between the two P type materials. As voltage is applied to the emitter, current will flow through the transistor and out the collector electrode. Without changing the voltage applied to the emitter, when a small amount of voltage is applied to the base electrode the amount of current passing through the collector electrode significantly increases. This way of increasing the current flow of a circuit by adjusting the voltage to the base electrode causes the transistor to act as an amplifier in the circuit. Think of applying a small amount of voltage to the base electrode as a small child who pulls a small lever to open the flood gates of a dam. A small voltage to the base can result in a large current increase through the collector.

T6B02 What electronic component allows current to flow in only one direction?
Answer: *Diode*.

Diodes only allow current to flow in one direction. Diodes consist of two layers: a P type material and a N type material side by side. The electrode associated with the P type material is called the anode and the electrode connected to the N type material is the cathode. If the anode and P type material are connected to the positive end of a power source and the cathode and N type material are connected to the negative end, current will flow through the diode freely. If the direction of current is reversed, the diode will stop the flow of current. The reason this happens is due to the properties of the interaction between the P type and N type materials at the point where they meet in the center of the diode called a PN junction. When the anode is connected to the positive end of a power source and the cathode to the negative end, the PN junction allows current to flow through the diode easily. When the current is reversed, the negative end of the powers source essentially pulls the positive characteristics of the P type material away from the N type material and the positive end of the power source pulls the negative characteristics of the N type away from the P type material. This creates a nonconducting region between the P type and N type materials called a depletion region. Current cannot flow through the depletion region. Therefore, diodes will only allow current to flow in one direction. The actual reason for this is much more technical than this. All you need to know for the exam is that when current is reversed, a diode stops the flow of electrical current.

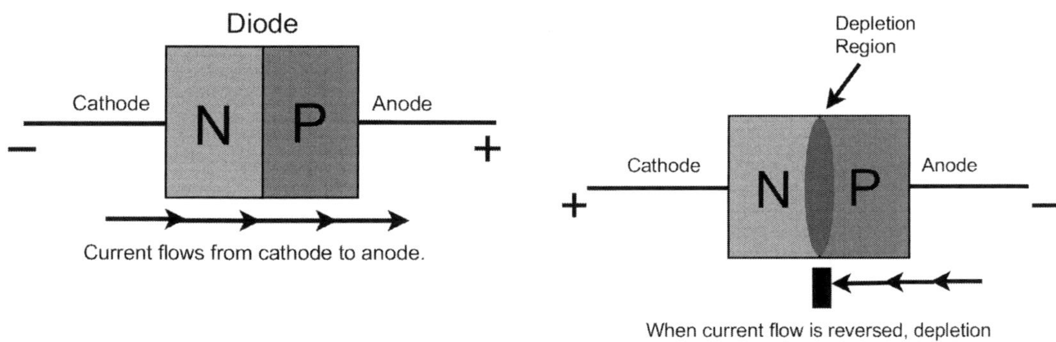

Current flows from cathode to anode.

When current flow is reversed, depletion region blocks current.

T6B03 Which of these components can be used as an electronic switch or amplifier?
Answer: *Transistor*.

We already talked about how transistors can be used as an amplifier in question one. Certain types of transistors can be used as a switch as well. An example of a transistor that can be used as a switch is a field effect transistor, or FET. The FET is a little different from the bipolar transistor we talked about in question 1. They still consist of three layers of semiconductor material and there are still three electrodes. The first difference is the names of the electrodes are different. The electrodes on a FET are the source, the drain, and the gate. We will use the same semiconductor material configuration we used in the bipolar transistor with a block of N type material in between two blocks of P type material. In a FET, the source and drain electrodes are connected to the opposite ends of the block of N type material. The gate electrode is shared with the two P type blocks on either side of the N type block. Current flows through the N type material from the source to the drain electrode. To limit or stop the current flow, a negative voltage is applied to the gate electrode connected to the two P type blocks. This causes a depletion region to form in the N type material choking the current flow. If the negative voltage applied to the gate is strong enough, the depletion region will cross the entire block of N type material stopping the current. The ability to stop the current flow allows the FET transistor to act as a switch.

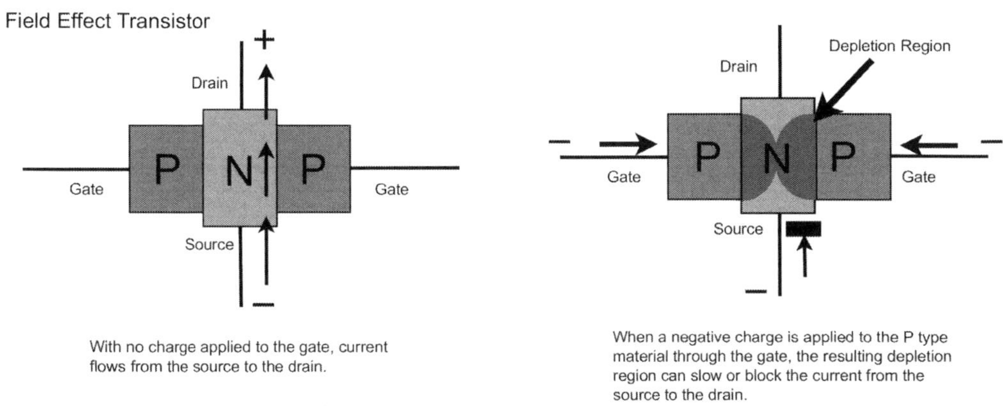

With no charge applied to the gate, current flows from the source to the drain.

When a negative charge is applied to the P type material through the gate, the resulting depletion region can slow or block the current from the source to the drain.

T6B04 Which of these components is made of three layers of semiconductor material?
Answer: *Bipolar junction transistor.*

You know that transistors have three layers of semiconductor material. The bipolar junction transistor is the only possible answer for this question that is a transistor. A transistor has three layers of semiconductor material.

T6B05 Which of the following electronic components can amplify signals?
Answer: *Transistor.*

As we discussed in question 1 of this lesson, one of the things a transistor can do is amplify signals.

T6B06 How is a semiconductor diode's cathode lead usually identified?
Answer: *With a stripe.*

Since diodes only allow electric current to flow in one direction, knowing which electrodes to hook to the positive and negative wires is important. Otherwise you run the risk of the diode blocking the current to the circuit. The cathode, or negative lead of the diode, is usually marked with a stripe.

T6B07 What does the abbreviation "LED" stand for?
Answer: *Light Emitting Diode.*

Light emitting diodes (LED), as common sense would have it, are diodes that emit light. They can produce a very bright light with relatively little power and are becoming more popular in items such as flashlights and traffic lights.

T6B08 What does the abbreviation "FET" stand for?
Answer: *Field Effect Transistor.*

We talked about the field effect transistor, or FET, in question 3. One of their functions is as a switch.

T6B09 What are the names of the two electrodes of a diode?
Answer: *Anode and cathode.*

The two electrodes of a diode are the anode (positive) and cathode(negative). Remember, the cathode is identified with a stripe.

T6B10 Which semiconductor component has an emitter electrode?
Answer: *Bipolar transistor.*

The three electrodes on a bipolar transistor are the emitter, collector, and base.

T6B11 Which semiconductor component has a gate electrode?
Answer: *Field effect transistor.*

The three electrodes on a FET are the source, drain, and gate.

T6B12 What is the term that describes a transistor's ability to amplify a signal?
Answer: *Gain.*

In dealing with transistors, the gain is the transistor's ability to amplify a signal. It is the answer that makes the most sense on the exam.

Quiz #20

(Find the answers on page 256)

1. _____ T6B01 What class of electronic components is capable of using a voltage or current signal to control current flow?
A. Capacitors
B. Inductors
C. Resistors
D. Transistors

2. _____ T6B02 What electronic component allows current to flow in only one direction?
A. Resistor
B. Fuse
C. Diode
D. Driven Element

3. _____ T6B03 Which of these components can be used as an electronic switch or amplifier?
A. Oscillator
B. Potentiometer
C. Transistor
D. Voltmeter

4. _____ T6B04 Which of these components is made of three layers of semiconductor material?
A. Alternator
B. Bipolar junction transistor
C. Triode
D. Pentagrid converter

5. _____ T6B05 Which of the following electronic components can amplify signals?
A. Transistor
B. Variable resistor
C. Electrolytic capacitor
D. Multi-cell battery

6. _____ T6B06 How is a semiconductor diode's cathode lead usually identified?
A. With the word "cathode"
B. With a stripe
C. With the letter "C"
D. All of these choices are correct

7. _____ T6B07 What does the abbreviation "LED" stand for?
A. Low Emission Diode
B. Light Emitting Diode
C. Liquid Emission Detector
D. Long Echo Delay

8. _____ T6B08 What does the abbreviation "FET" stand for?
A. Field Effect Transistor
B. Fast Electron Transistor
C. Free Electron Transition
D. Field Emission Thickness

9. _____ T6B09 What are the names of the two electrodes of a diode?
A. Plus and minus
B. Source and drain
C. Anode and cathode
D. Gate and base

10. _____ T6B10 Which semiconductor component has an emitter electrode?
A. Bipolar transistor
B. Field effect transistor
C. Silicon diode
D. Bridge rectifier

11. _____ T6B11 Which semiconductor component has a gate electrode?
A. Bipolar transistor
B. Field effect transistor
C. Silicon diode
D. Bridge rectifier

12. _____ T6B12 What is the term that describes a transistor's ability to amplify a signal?
A. Gain
B. Forward resistance
C. Forward voltage drop
D. On resistance

Lesson 20 Study Sheet

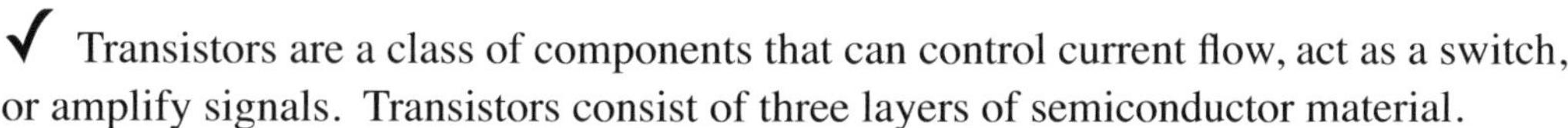

✓ Transistors are a class of components that can control current flow, act as a switch, or amplify signals. Transistors consist of three layers of semiconductor material.

✓ Diodes only allow current to flow in one direction.

✓ The three electrodes of a bipolar transistor are the emitter, collector, and base.

✓ The three electrodes of a field effect transistor (FET) are the source, drain, and gate.

✓ The two electrodes of a diode are the anode and cathode. The cathode is marked with a stripe.

Notes:

Lesson 21
The T6C Section
Circuit Diagrams

T6C01 What is the name for standardized representations of components in an electrical wiring diagram?
Answer: *Schematic symbols.*

Schematic symbols are recognized widely graphic representations of electrical components. These schematic symbols are used to with schematic diagrams. Schematic diagrams are the blue prints of a circuit. Schematic diagrams use schematic symbols to show how the various components are interconnected in a circuit. Schematic diagrams and symbols are in no way to scale. They are simply standardized representations used for the purpose of illustrating the interconnectivity of components in a circuit.

[Use Figure T1 for questions 2 through 5].

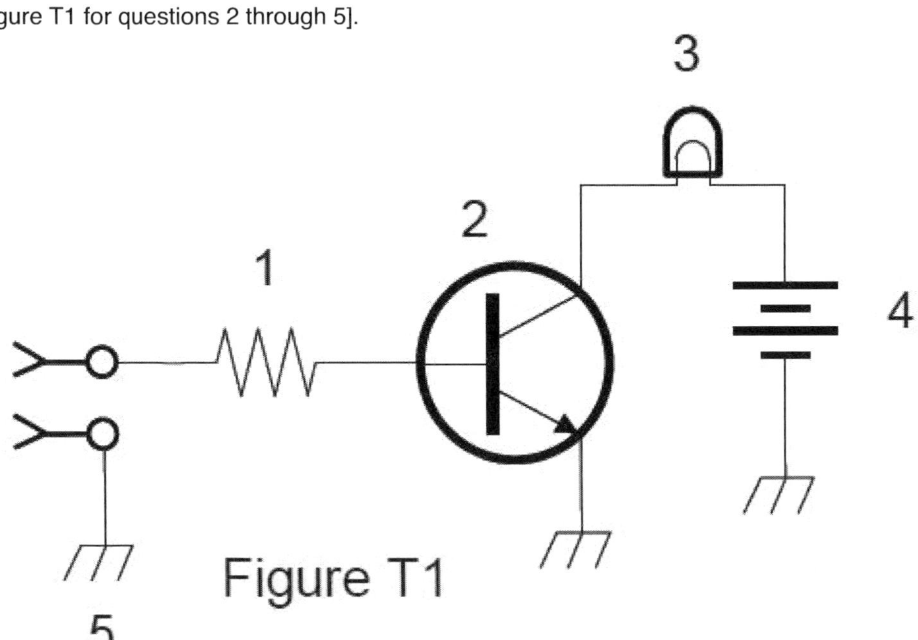

Figure T1

T6C02 What is component 1 in figure T1?
Answer: *Resistor*.

A resistor is a component that opposes the flow of current. The schematic symbol for a resistor is a series of zigzags and looks like this:

Resistor

T6C03 What is component 2 in figure T1?
Answer: *Transistor*.

A transistor is a type of semiconductor consisting of three layers of semiconductor material. They are used as switches and amplifiers. The schematic symbol for a transistor looks like this:

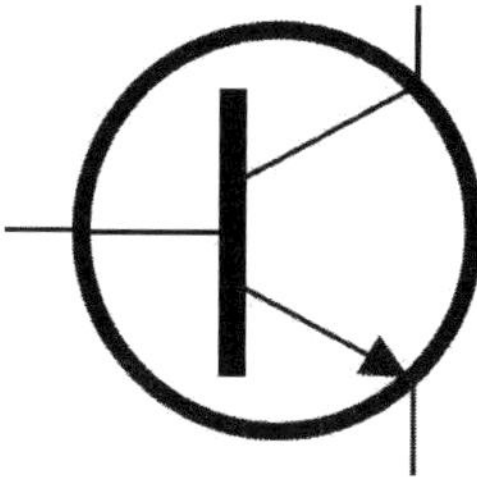

Transistor

T6C04 What is component 3 in figure T1?
Answer: *Lamp*.

The schematic symbol for a lamp looks a little like a lightbulb.

Lamp

T6C05 What is component 4 in figure T1?
Answer: *Battery*.

A battery is a direct current (DC) power source. The schematic symbol for a battery consists of a series of alternating short and long parallel lines.

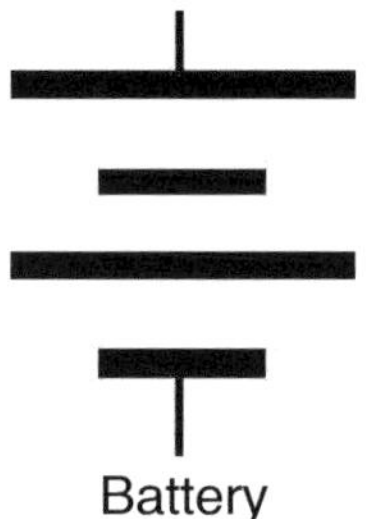

[Use Figure T2 for questions 6 through 9.]

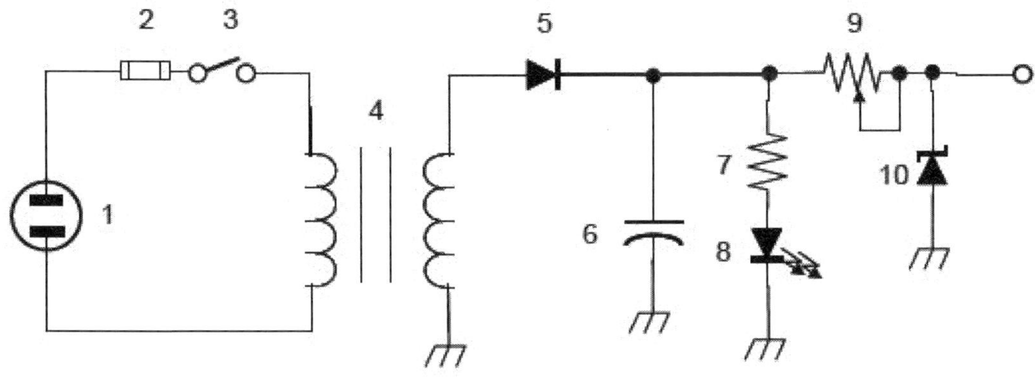

Figure T2

T6C06 What is component 6 in figure T2?
Answer: *Capacitor*.

A capacitor is made of two or more plates separated by an insulator. With this in mind, the schematic symbol for a capacitor make sense. It looks like two surfaces, except one is a little bent.

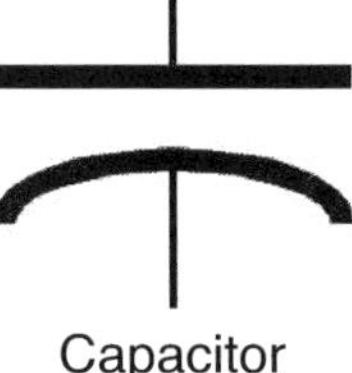

T6C07 What is component 8 in figure T2?
Answer: *Light emitting diode.*

The schematic symbol for a light emitting diode, or LED, consists of the schematic symbol of a diode with lightning bolts coming out of it. The lighting bolts help tell you that the diode is emitting something. In this case, the diode is emitting light.

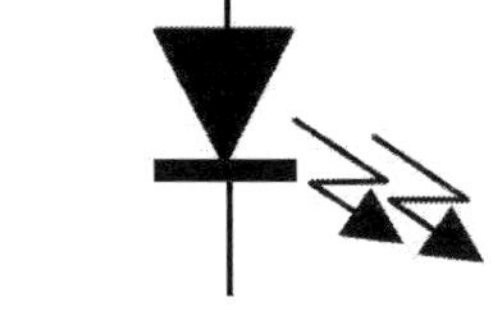
Light Emitting Diode (LED)

T6C08 What is component 9 in figure T2?
Answer: Variable resistor.

A variable resistor is a component which is capable of adjusting the amount of resistance in a circuit. The schematic symbol for a variable resistor consists of the schematic symbol of a resistor with an arrow pointing at its center. The arrow lets you know that the amount of resistance is adjustable.

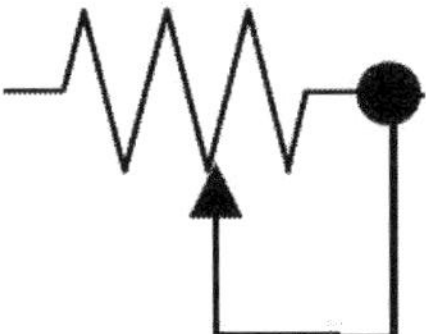
Variable Resistor

T6C09 What is component 4 in figure T2?
Answer: *Transformer.*

A transformer are essentially two inductors which, when they work together, are capable of changing the voltage in a circuit. Transformers are often used by overseas travelers who need to plug an electronic device rated for 120 volts (the U.S. standard) into a foreign outlet with a greater outlet voltage. The transformer reduces the voltage from the outlet to 120 volts and avoids damaging the electronic device or avoids starting a fire. Remember that an inductor usually consists of a coil of wire. The schematic for a transformer looks like two coils of wire.

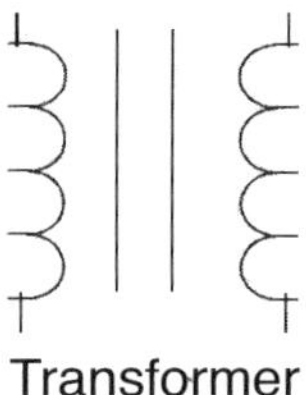
Transformer

[Use Figure T3 for questions 10 and 11.]

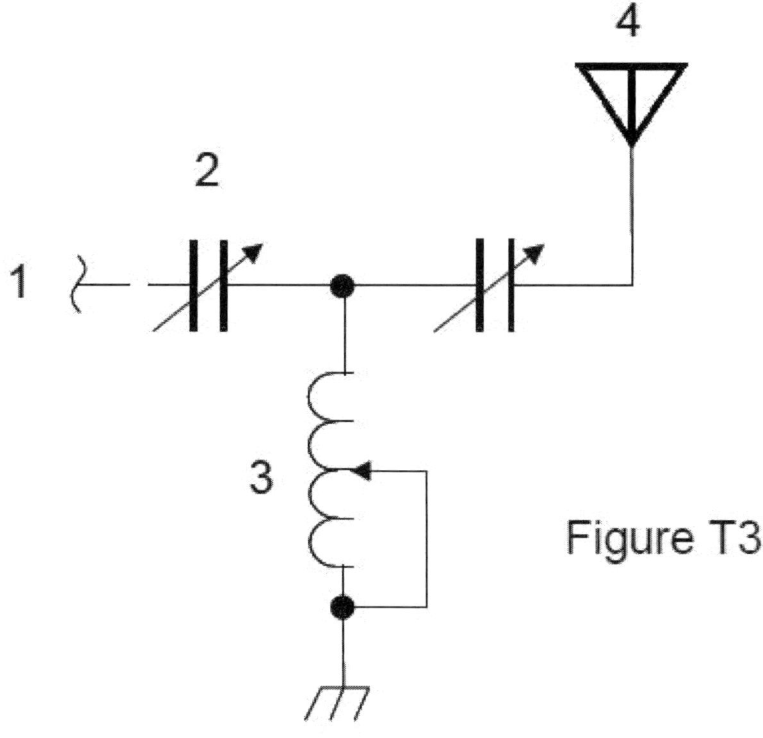

Figure T3

T6C10 What is component 3 in figure T3?
Answer: *Variable inductor.*

Like variable resistors, some inductors are capable of adjusting the level of inductance they add to a circuit. These are called variable inductors. The schematic for an inductor looks something like a coil of wire. For the variable inductor, the schematic symbol has an arrow pointing to the center of the inductor letting you know that the inductance can be adjusted.

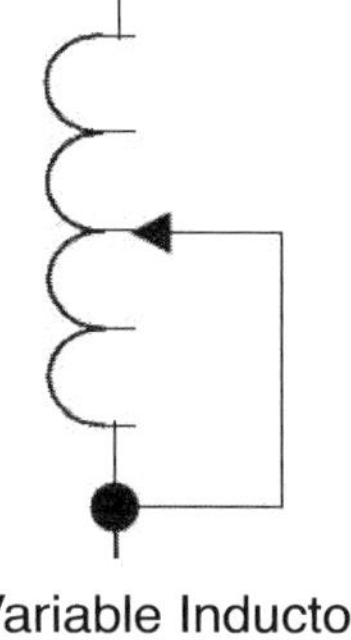

Variable Inductor

T6C11 What is component 4 in figure T3?
Answer: *Antenna.*

The schematic symbol for an antenna sort of looks like an antenna. It is an inverted triangle on a stick.

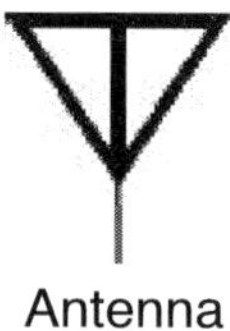

Antenna

T6C12 What do the symbols on an electrical circuit schematic diagram represent?
Answer: *Electrical components.*

All the schematic symbols we went over in this lesson represent electrical components on a schematic diagram. These are standardized symbols which allow them to be widely recognized.

T6C13 Which of the following is accurately represented in electrical circuit schematic diagrams?
Answer: *The way components are interconnected.*

Schematic diagrams are not to scale. They simply show how components in a circuit are interconnected.

Quiz #21

(Find the answers on page 256)

1. _____ T6C01 What is the name for standardized representations of components in an electrical wiring diagram?
A. Electrical depictions
B. Grey sketch
C. Schematic symbols
D. Component callouts

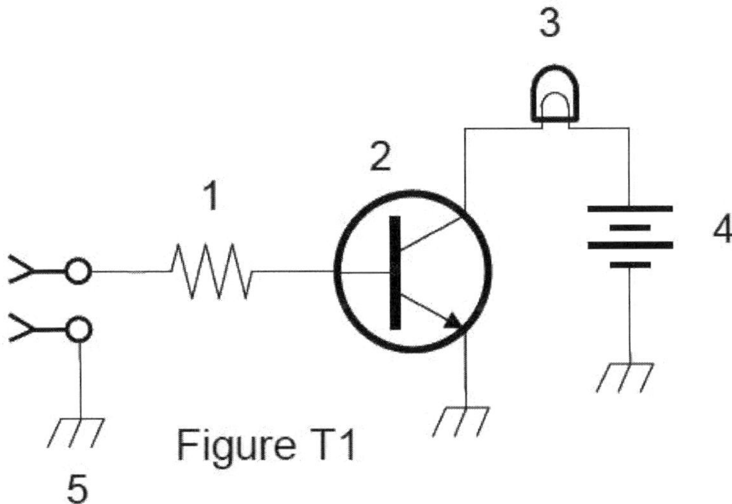

Figure T1

2. _____ T6C02 What is component 1 in figure T1?
A. Resistor
B. Transistor
C. Battery
D. Connector

3. _____ T6C03 What is component 2 in figure T1?
A. Resistor
B. Transistor
C. Indicator lamp
D. Connector

4. _____ T6C04 What is component 3 in figure T1?
A. Resistor
B. Transistor
C. Lamp
D. Ground symbol

5. _____ T6C05 What is component 4 in figure T1?
A. Resistor
B. Transistor
C. Battery
D. Ground symbol

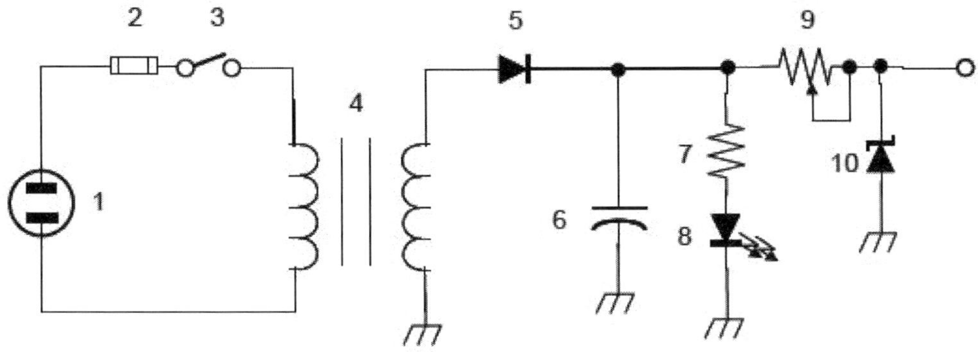

Figure T2

6. _____ T6C06 What is component 6 in figure T2?
A. Resistor
B. Capacitor
C. Regulator IC
D. Transistor

7. _____ T6C07 What is component 8 in figure T2?
A. Resistor
B. Inductor
C. Regulator IC
D. Light emitting diode

8. _____ T6C08 What is component 9 in figure T2?
A. Variable capacitor
B. Variable inductor
C. Variable resistor
D. Variable transformer

9. _____ T6C09 What is component 4 in figure T2?
A. Variable inductor
B. Double-pole switch
C. Potentiometer
D. Transformer

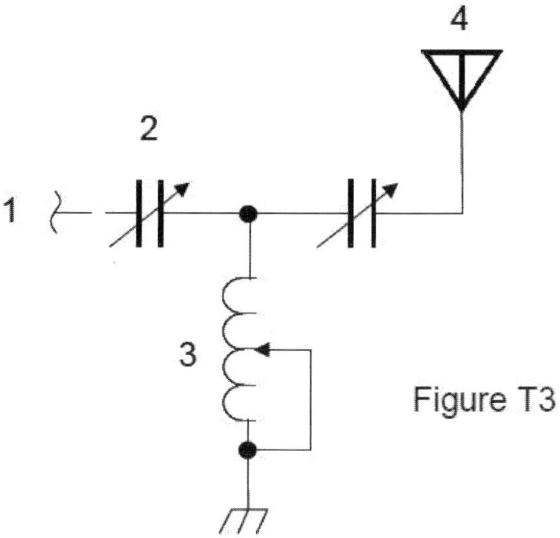

Figure T3

10. _____ T6C10 What is component 3 in figure T3?
A. Connector
B. Meter
C. Variable capacitor
D. Variable inductor

11. _____ T6C11 What is component 4 in figure T3?
A. Antenna
B. Transmitter
C. Dummy load
D. Ground

12. _____ T6C12 What do the symbols on an electrical circuit schematic diagram represent?
A. Electrical components
B. Logic states
C. Digital codes
D. Traffic nodes

13. _____ T6C13 Which of the following is accurately represented in electrical circuit schematic diagrams?
A. Wire lengths
B. Physical appearance of components
C. The way components are interconnected
D. All of these choices are correct

Lesson 21 Study Sheet

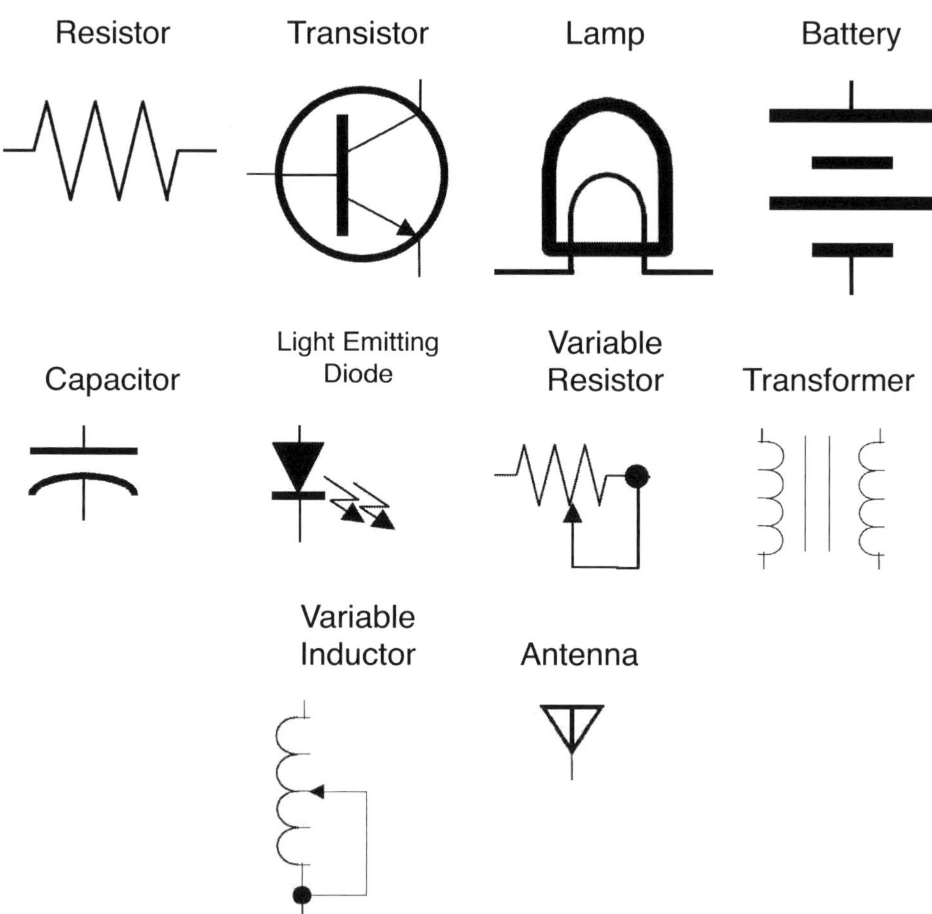

Notes:

Lesson 22
The T2D Section
Component Functions

T6D01 Which of the following devices or circuits changes an alternating current into a varying direct current signal?
Answer: *Rectifier.*

A rectifier is essentially a diode. Rectification is the process of turning alternating current (AC) into direct current (DC). Diodes only allow current to flow in one direction. If an alternating current is applied to a diode, it will allow current to flow in one of the directions, but will block the AC current when the current flows the opposite direction. The result turns the alternating current into varying direct current. The rectifier will only allow the current to flow in the direction. Since AC current reverses itself back and forth constantly, the current will flow in the direction the rectifier will allow only half the time. Instead of a smooth DC flow of current, the result is a DC current that pulses on and off, or in other words, varies. For this question you need to know that rectifiers and diodes are essentially the same and that a rectifier will change an AC current into varying DC current.

T6D02 What best describes a relay?
Answer: *A switch controlled by an electromagnet.*

A relay is a switch that is controlled electronically through an electromagnet. A relay is simply a type of remote switch.

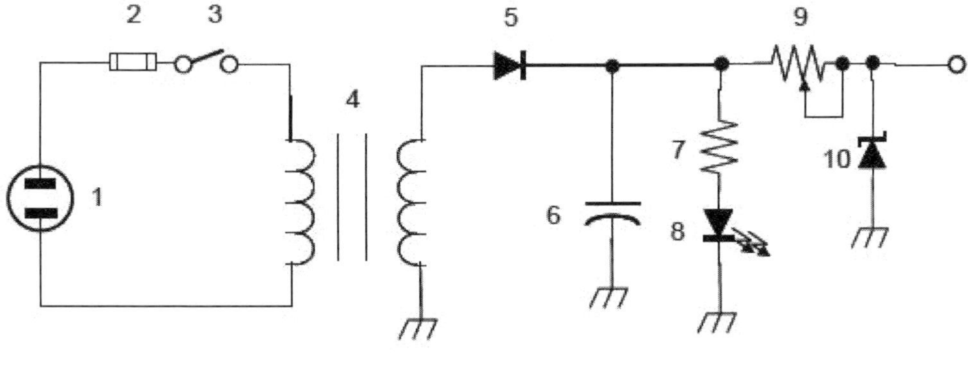

Figure T2

T6D03 What type of switch is represented by item 3 in figure T2?
Answer: *Single-pole single-throw.*

A single-pole single-throw switch is simply an on and off switch. If you look at the schematic symbol it will give you a good idea of its function. With a single-pole single-throw switch the switch can either let current flow through the circuit or short the circuit cutting off the flow.

T6D04 Which of the following can be used to display signal strength on a numeric scale?
Answer: *Meter.*

Another question not to over think. A voltmeter, ammeter, wattmeter, and S meter all display their respective measurements on a numeric scale and all are types of meters.

T6D05 What type of circuit controls the amount of voltage from a power supply?
Answer: *Regulator.*

A regulator controls, or regulates, the amount of voltage from a power supply. A regulator is a type of circuit that ensures the voltage coming from a power supply fluctuates as little as possible and prevents performance issues or damage to your equipment.

T6D06 What component is commonly used to change 120V AC house current to a lower AC voltage for other uses?
Answer: *Transformer.*

A transformer changes higher voltages to lower voltages required by some electric devices. Not every thing you plug into an outlet can handle 120 volts. A transformer will reduce the voltage to a level your equipment can handle. This prevents overloading, destruction of your equipment, or a fire.

T6D07 Which of the following is commonly used as a visual indicator?
Answer: *LED.*

LEDs, or light emitting diodes, are often used as a visual indicator. That little red light that indicates your TV is on is probably an LED.

T6D08 Which of the following is used together with an inductor to make a tuned circuit?
Answer: *Capacitor.*

The capacitor/inductor combination is what makes the amateur radio world go round! Together they act like a pendulum on a clock, kicking energy back and forth causing the circuit to oscillate. A capacitor stores energy in an electric field and an inductor stores energy in a magnetic field. When the current to a capacitor is turned off, the capacitor does not release its stored energy all at once. The capacitor will release it over a short period of time. In the same way, once the current is cut off to an inductor, it releases the energy stored in its magnetic field over a short period of time. If a capacitor and inductor are connected in a circuit, as the capacitor releases its stored energy it is picked up by the inductor, building up the inductor's magnetic field and storing energy. Once the capacitor has released all of its stored energy there is no more current going to the inductor. With no more current going to the inductor, it will start to release the energy stored in its magnetic field and send it back to the capacitor putting energy back into the capacitor's electric field. The capacitor and inductor will continue this exchange of energy until all the energy in the circuit is dispersed through imperfections and resistance in the wiring.

T6D09 What is the name of a device that combines several semiconductors and other components into one package?
Answer: *Integrated circuit.*

An integrated circuit (IC) integrates the functions of several electronic components and semiconductors into a neat little package often on a microchip. Integrated circuits are big space savers and can allow a lot functionality in a small space.

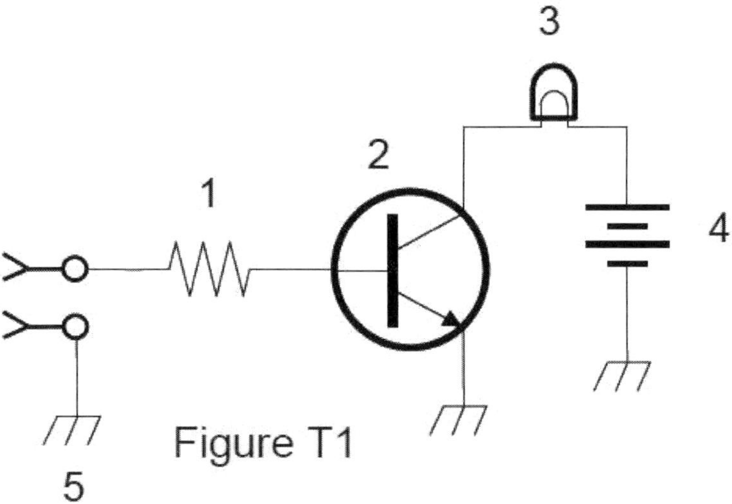

Figure T1

T6D10 What is the function of component 2 in Figure T1?
Answer: *Control the flow of current.*

Component #2 in Figure T1 is a transistor. The transistor can act as an amplifier or switch to control the current flow in the circuit

T6D11 Which of the following is a common use of coaxial cable?
Answer: *Carry RF signals between a radio and antenna.*

Coaxial cable (or coax) is a type of cable that carries the signal from your transceiver to your antenna. Coax comes in handy because it is easily water proofed and is very low maintenance compared to other types of cable or feedlines designed to carry signals. Coaxial cable is also the same type of wiring cable television companies use to pipe cable television into your house.

Quiz #22

(Find the answers on page 256)

1. _____ T6D01 Which of the following devices or circuits changes an alternating current into a varying direct current signal?
A. Transformer
B. Rectifier
C. Amplifier
D. Reflector

2. _____ T6D02 What best describes a relay?
A. A switch controlled by an electromagnet
B. A current controlled amplifier
C. An optical sensor
D. A pass transistor

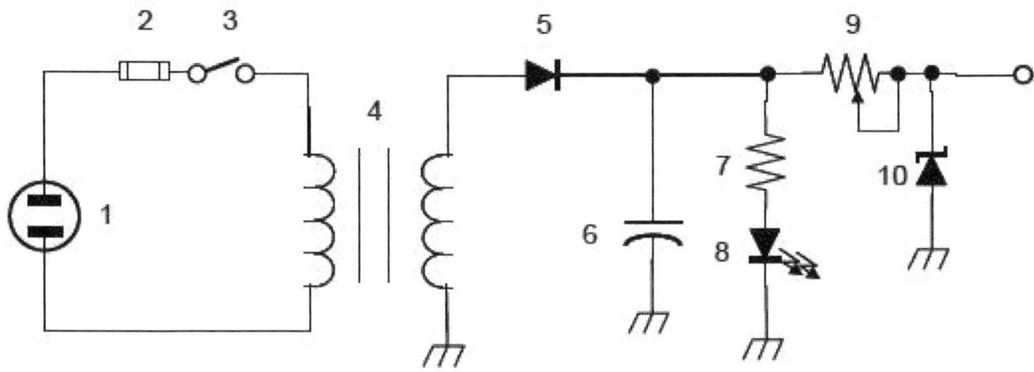

Figure T2

3. _____ T6D03 What type of switch is represented by item 3 in figure T2?
A. Single-pole single-throw
B. Single-pole double-throw
C. Double-pole single-throw
D. Double-pole double-throw

4. _____ T6D04 Which of the following can be used to display signal strength on a numeric scale?
A. Potentiometer
B. Transistor
C. Meter
D. Relay

5. _____ T6D05 What type of circuit controls the amount of voltage from a power supply?
A. Regulator
B. Oscillator
C. Filter
D. Phase inverter

6. _____ T6D06 What component is commonly used to change 120V AC house current to a lower AC voltage for other uses?
A. Variable capacitor
B. Transformer
C. Transistor
D. Diode

7. _____ T6D07 Which of the following is commonly used as a visual indicator?
A. LED
B. FET
C. Zener diode
D. Bipolar transistor

8. _____ T6D08 Which of the following is used together with an inductor to make a tuned circuit?
A. Resistor
B. Zener diode
C. Potentiometer
D. Capacitor

9. _____ T6D09 What is the name of a device that combines several semiconductors and other components into one package?
A. Transducer
B. Multi-pole relay
C. Integrated circuit
D. Transformer

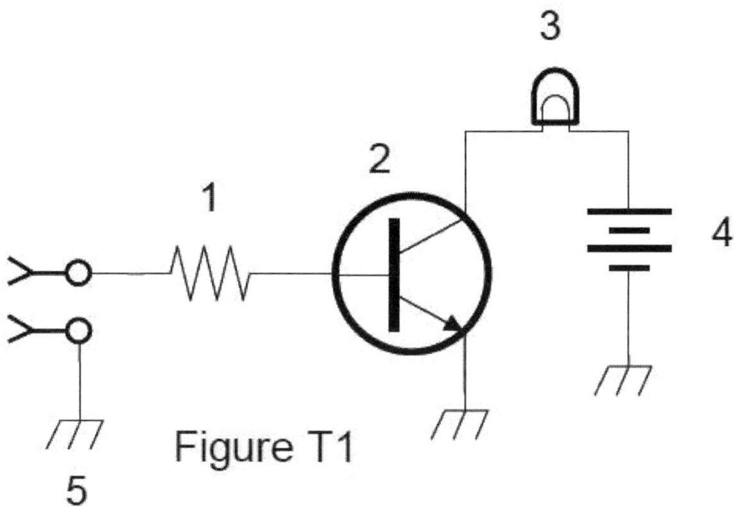

Figure T1

10. _____ T6D10 What is the function of component 2 in Figure T1?
A. Give off light when current flows through it
B. Supply electrical energy
C. Control the flow of current
D. Convert electrical energy into radio waves

11. _____ T6D11 Which of the following is a common use of coaxial cable?
A. Carry dc power from a vehicle battery to a mobile radio
B. Carry RF signals between a radio and antenna
C. Secure masts, tubing, and other cylindrical objects on towers
D. Connect data signals from a TNC to a computer

Lesson 22 Study Sheet

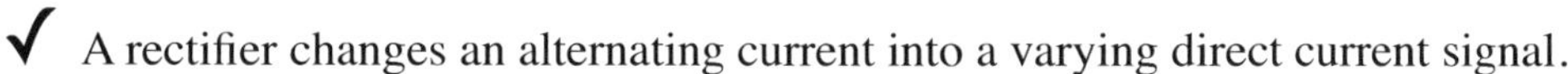

✓ A rectifier changes an alternating current into a varying direct current signal.

✓ A relay is a switch controlled by an electromagnet.

✓ A regulator controls the amount of voltage from a power supply.

✓ A capacitor works together with an inductor to make a tuned circuit.

Notes:

Lesson 23
The T7A Section
Station Radios

T7A01 What is the function of a product detector?
Answer: *Detect CW and SSB signals.*

A product detector is a circuit in a receiver which demodulates CW and SSB signals. What this basically means is it converts the radio signal into audio tones you can understand. A product detector works with a beat frequency oscillator (BFO). The BFO combines the incoming radio frequency with a second frequency. The product of the two frequencies creates a third frequency called an intermediate frequency (IF). This IF is in the audio frequency range. Product detectors are really only used for CW and SSB. They are not used with FM. To sum up, a product detector transforms the frequency of incoming CW and SSB signals to a frequency the human ear can hear.

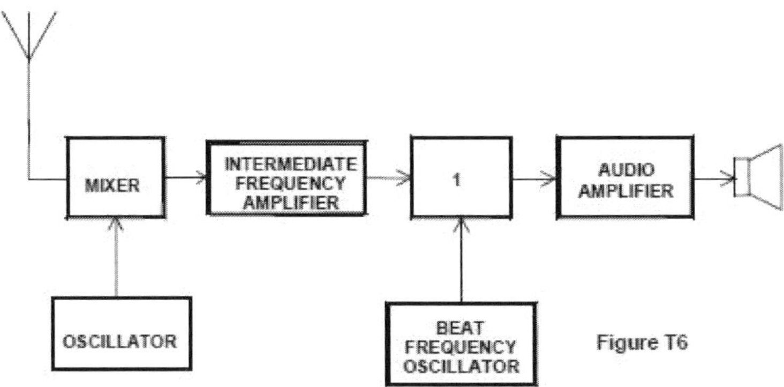

Figure T6

T7A02 What type of receiver is shown in Figure T6?
Answer: *Single-conversion superheterodyne.*

If you break the word "heterodyne" into its verbal roots, it means "other power." This is sort of a hint to how to identify the block diagram in Figure T6. A superheterodyne receiver is characterized by a mixer circuit working together with an oscillator which produces and intermediate frequency for the receiver to process. As the signal comes in from the antenna, the mixer mixes the incoming frequency with a signal produced by an oscillator. An oscillator is essentially a tiny transmitter which produces a signal at a specific frequency. The combination of these signals produces a new intermediate frequency (IF) signal. This IF signal can be filtered and is converted to audio frequency ranges by, perhaps, a product detector and BFO (if it is a CW or SSB signal). What identifies Figure T6 as a superheterodyne receiver is the mixer and oscillator combination. What identifies it as a single-conversion superheterodyne receiver is there is only one mixer/oscillator combination. Double and triple-conversion superheterodyne receivers have two or three, respectively.

T7A03 What is the function of a mixer in a superheterodyne receiver?
Answer: *To shift the incoming signal to an intermediate frequency.*

The mixer in a superheterodyne receiver receives the incoming radio signal from the antenna plus a second signal of a different frequency from an oscillator. The combination of these two frequencies produces a new third frequency, either the sum or the difference of the two mixed frequencies, called an intermediate frequency (IF). For instance, if the incoming radio signal is 7 MHz and the frequency produced by the oscillator is 5 MHz, the IF will be 12 MHz or 2 MHz.

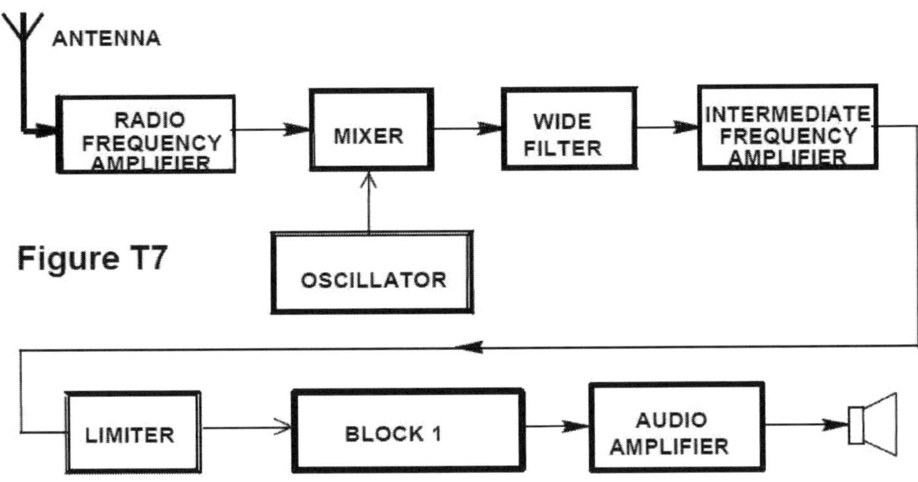

Figure T7

T7A04 What circuit is pictured in Figure T7, if block 1 is a frequency discriminator?
Answer: *An FM receiver.*

A frequency discriminator performs the same function for an FM receiver as a product detector does for CW and SSB signals. It converts the incoming radio frequency (in this case an intermediate frequency) to audio frequencies you can listen to and understand. When you see "discriminator," think FM.

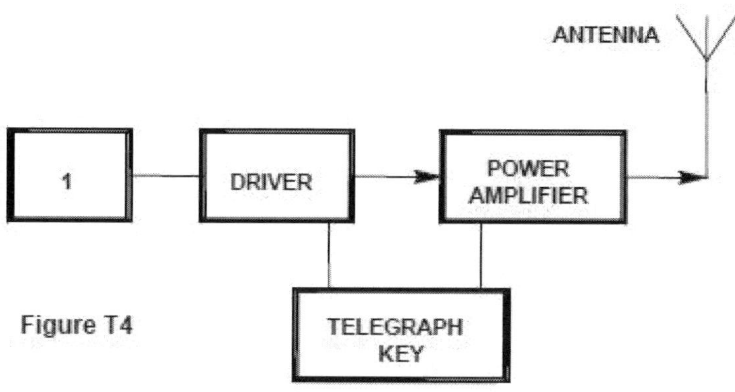

Figure T4

T7A05 What is the function of block 1 if figure T4 is a simple CW transmitter?
Answer: *Oscillator.*

Block 1 is an oscillator. The frequency of the oscillator for the CW transmitter figure T4 will be the frequency of the transmitters signal when the telegraph key is pressed. The driver and power amplifiers are both types of amplifiers which increase the strength of the signal. Every transmitter needs an oscillator to help determine the transmitting frequency. The only thing missing from this transmitter block diagram is an oscillator.

T7A06 What device takes the output of a low-powered 28 MHz SSB exciter and produces a 222 MHz output signal?
Answer: *Transverter.*

Transverters convert one frequency to another frequency for transmission. In this question, a transverter will convert the radio signal frequency from a 10 meter transmitter to frequencies in the 1.25 meter band.

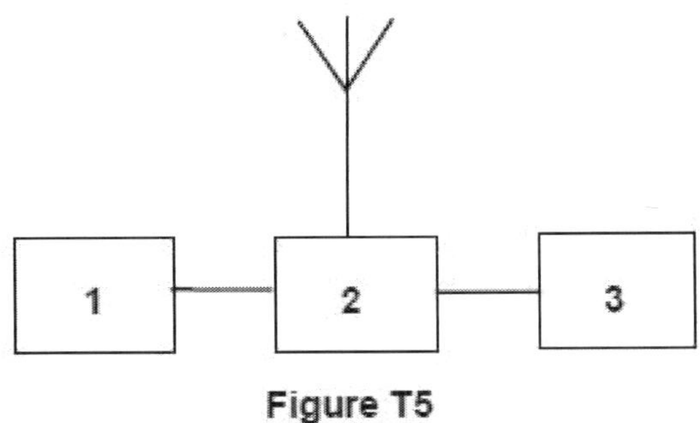

Figure T5

T7A07 If figure T5 represents a transceiver in which block 1 is the transmitter portion and block 3 is the receiver portion, what is the function of block 2?
Answer: *A transmit-receive switch.*

Unfortunately, if your transmitter and receiver are sharing an antenna, you cannot transmit and receive at the same time. If you try both at the same time, when you transmit you will overload your receiver and that will be it for your hobby until you get it fixed. The transmit-receive switch provides harmony to the transmitter/receiver relationship. With a manual transmit-receive switch, the transmitter remains disconnected from the antenna until you flip the switch to disconnect the receiver's antenna connection. In a transceiver (a combined transmitter and receiver), the transmit-receive switch is automatic. When you key the transmitter the receiver is disconnected from the antenna.

T7A08 Which of the following circuits combines a speech signal and an RF carrier?
Answer: *Modulator.*

An RF carrier is a radio signal with no information on it. It is just some radio waves at a certain frequency that do not say anything. To put speech information on that RF carrier you need to modulate the carrier wave in some form. To do this you use a modulator. A modulator takes signals from the transmitter's microphone and uses them to modify the carrier wave so that it will produce a radio signal that carries information, like the sound of your voice.

T7A09 Which of the following devices is most useful for VHF weak-signal communication?
Answer: *A multi-mode VHF transceiver.*

Of the possible choices on the exam, this answer is the best. For VHF weak signal communication you need something more than just FM. FM is not a good weak signal mode. SSB and CW are much better weak signal modes. If you had a VHF transceiver that handled all these modes (otherwise known as a multi-mode transceiver), you would have a useful device for weak signal communication. It also may help to eliminate some of the other possible answers. A quarter-wave vertical antenna is a bit too short and the wrong polarization for weak signal operation. An omnidirectional antenna is not focused enough, a directional antenna is better. You should probably forget any kind of weak signal operation with a mobile VHF FM transceiver. Mobile radios are not great for weak signal operation and FM is probably the worst mode for weak signals.

T7A10 What device increases the low-power output from a handheld transceiver?
Answer: *An RF power amplifier.*

When you need to increase the RF power output from a handheld or any type of transceiver, you use an RF power amplifier. Another question not to over think.

T7A11 Which of the following circuits demodulates FM signals?
Answer: *Discriminator.*

When you see "discriminator" think FM. The discriminator in an FM receiver converts the signal frequency to audio frequencies so you can hear and understand the information received.

T7A12 Which term describes the ability of a receiver to discriminate between multiple signals?
Answer: *Selectivity.*

Selectivity is the ability of a receiver to discriminate between multiple signals. A receiver's selectivity is a measure of how well it can pick out the signal you want from all the other RF noise the antenna is receiving. Filters help improve a receiver's selectivity. Do not get this confused with sensitivity! A receiver's sensitivity is determined by how well it picks up weak signals. Selectivity chooses the preferred signal out of multiple signals.

T7A13 Where is an RF preamplifier installed?
Answer: *Between the antenna and receiver.*

An RF preamplifier increases the strength of all the signals the antenna picks up before the signals are sent to the receiver. RF preamplifiers help improve the sensitivity of a receiver and its ability to pick up weak signals.

Quiz #23

(Find the answers on page 256)

1. _____ T7A01 What is the function of a product detector?
A. Detect phase modulated signals
B. Demodulate FM signals
C. Detect CW and SSB signals
D. Combine speech and RF signals

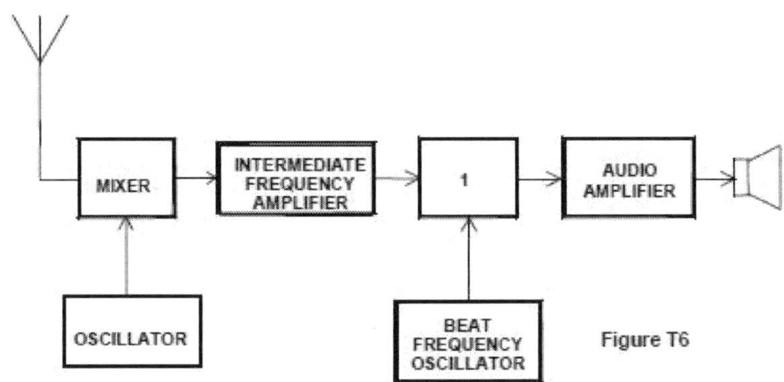

Figure T6

2. _____ T7A02 What type of receiver is shown in Figure T6?
A. Direct conversion
B. Super-regenerative
C. Single-conversion superheterodyne
D. Dual-conversion superheterodyne

3. _____ T7A03 What is the function of a mixer in a superheterodyne receiver?
A. To reject signals outside of the desired passband
B. To combine signals from several stations together
C. To shift the incoming signal to an intermediate frequency
D. To connect the receiver with an auxiliary device, such as a TNC

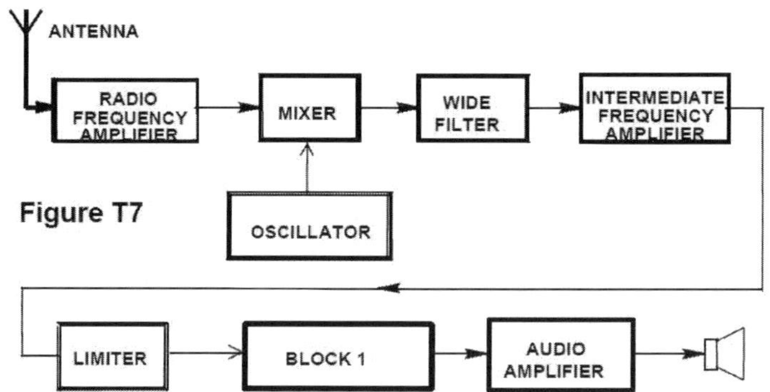

Figure T7

4. _____ T7A04 What circuit is pictured in Figure T7, if block 1 is a frequency discriminator?
A. A double-conversion receiver
B. A regenerative receiver
C. A superheterodyne receiver
D. An FM receiver

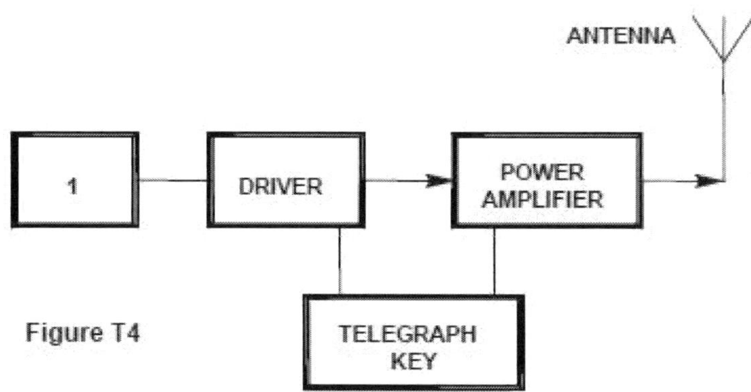

Figure T4

5. _____ T7A05 What is the function of block 1 if figure T4 is a simple CW transmitter?
A. Reactance modulator
B. Product detector
C. Low-pass filter
D. Oscillator

6. _____ T7A06 What device takes the output of a low-powered 28 MHz SSB exciter and produces a 222 MHz output signal?
A. High-pass filter
B. Low-pass filter
C. Transverter
D. Phase converter

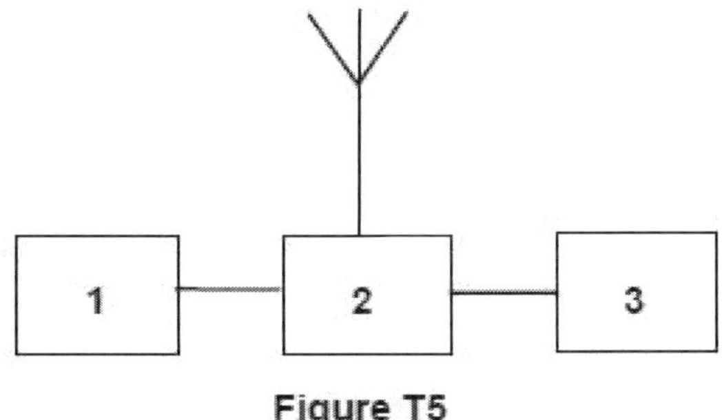

Figure T5

7. _____ T7A07 If figure T5 represents a transceiver in which block 1 is the transmitter portion and block 3 is the receiver portion, what is the function of block 2?
A. A balanced modulator
B. A transmit-receive switch
C. A power amplifier
D. A high-pass filter

8. _____ T7A08 Which of the following circuits combines a speech signal and an RF carrier?
A. Beat frequency oscillator
B. Discriminator
C. Modulator
D. Noise blanker

9. _____ T7A09 Which of the following devices is most useful for VHF weak-signal communication?
A. A quarter-wave vertical antenna
B. A multi-mode VHF transceiver
C. An omni-directional antenna
D. A mobile VHF FM transceiver

10. _____ T7A10 What device increases the low-power output from a handheld transceiver?
A. A voltage divider
B. An RF power amplifier
C. An impedance network
D. A voltage regulator

11. _____ T7A11 Which of the following circuits demodulates FM signals?
A. Limiter
B. Discriminator
C. Product detector
D. Phase inverter

12. _____ T7A12 Which term describes the ability of a receiver to discriminate between multiple signals?
A. Tuning rate
B. Sensitivity
C. Selectivity
D. Noise floor

13. _____ T7A13 Where is an RF preamplifier installed?
A. Between the antenna and receiver
B. At the output of the transmitter's power amplifier
C. Between a transmitter and antenna tuner
D. At the receiver's audio output

Lesson 23 Study Sheet

✓ A product detector detects CW and SSB signals and converts them to audio frequencies.

✓ A frequency discriminator demodulates FM signals and converts them to audio.

✓ Modulators are circuits that combine a speech signal and an RF carrier.

✓ A transverter takes the output of a low-powered 28 MHz SSB exciter and produces a 222 MHz output signal.

Notes:

Lesson 24
The T7B Section
Common Transmitter and Receiver Problems

T7B01 What can you do if you are told your FM handheld or mobile transceiver is over deviating?
Answer: *Talk farther away from the microphone.*

Speaking too loudly or too close to a microphone can cause your transmitter to send out a disorganized signal that can deviate from the intended frequency. Speaking loudly or too close to the microphone is the FM equivalent of throwing paint at a canvas. It's sloppy and not all the paint stays on the canvas. Try holding the microphone a little farther away from your mouth and speak in a calm voice. This will help.

T7B02 What is meant by fundamental overload in reference to a receiver?
Answer: *Interference caused by very strong signals.*

Very strong signals may overwhelm your receiver's ability to filter out the noise and focus on the signal you are looking for. Overload can occur if your antenna is close to another transmitting antenna. The receiver is simply picking up very strong signals which overload the receiver's ability to be selective.

T7B03 Which of the following may be a cause of radio frequency interference?
Answer: *Fundamental overload, harmonics, and spurious emissions.*

This is an "all of the above" question on the exam. Fundamental overload, harmonics, and spurious emissions can all cause interference. Fundamental overloads are strong signals that interfere with a receiver. Harmonics are an unwanted byproduct of transmissions that can be minimized by filters. Spurious emissions are also unwanted signals which can interfere with electrical equipment. If you or your neighbors are experiencing interference, make sure your equipment is in good working order and the problem is not you. Check to make sure that all your station connections are solid. Adding filters to you equipment or nearby electric devices may help as well.

T7B04 What is the most likely cause of interference to a non-cordless telephone from a nearby transmitter?
Answer: *The telephone is inadvertently acting as a radio receiver.*

Some phones, especially older phones, may pick up strong radio signals and inadvertently act as a receiver. Installing a RF filter at the telephone will usually help solve the problem.

T7B05 What is a logical first step when attempting to cure a radio frequency interference problem in a nearby telephone?
Answer: *Install an RF filter at the telephone.*

You are not going to have too much success fixing the problem at your transmitter. The best way to stop your transmissions from interfering with a telephone is to block all the RF from getting to the telephone. The easiest way to do that is to install an RF filter at the telephone.

T7B06 What should you do first if someone tells you that your station's transmissions are interfering with their radio or TV reception?
Answer: *Make sure that your station is functioning properly and that it does not cause interference to your own television.*

A problem with being a ham operator is that when people are looking for the source of interference in their television or radio, the first thing they see is the giant antenna in your backyard. If one of your neighbors approaches you about causing interference, the first thing you should do is check to make sure your station is working properly and your equipment and antenna connections are solid. Then try doing a test transmission to see if you are experiencing interference on your own television. Whenever this happens, always be nice, be courteous, and be polite. Hostility gets you nowhere.

T7B07 Which of the following may be useful in correcting a radio frequency interference problem?
Answer: *Snap-on ferrite chokes, low-pass and high-pass filters, and band-reject and band-pass filters.*

Another "all of the above" question on the exam. Ferrite chokes, high and low-pass filters, and band-reject filters all can be used to prevent radio frequency interference (RFI) problems.

T7B08 What should you do if a "Part 15" device in your neighbor's home is causing harmful interference to your amateur station?
Answer: *Work with your neighbor to identify the offending device, politely inform your neighbor about the rules that require him to stop using the device if it causes interference, and check your station and make sure it meets the standards of good amateur practice.*

Yet another "all of the above" question. An FCC "Part 15" device is any electronic device that produces RF signals and doesn't require a license to operate. Microwave ovens, burglar alarms, garage door openers, and TV remote controls are examples of "Part 15" devices. The first thing you need to do if you are experiencing interference is make sure you are not the problem and your station is in good working order. Next approach your neighbor, politely, about the problem and assist them in locating and correcting the device causing interference. Always be polite and helpful.

T7B09 What could be happening if another operator reports a variable high-pitched whine on the audio from your mobile transmitter?
Answer: *Noise on the vehicle's electrical system is being transmitted along with your speech audio.*

We talked about alternator interference causing a whine in your receiver. Your vehicle's alternator can also cause interference in you transmitted signal. If this is a problem, making sure you have a good solid ground and installing filters at the transceiver may help.

T7B10 What might be the problem if you receive a report that your audio signal through the repeater is distorted or unintelligible?
Answer: *Your transmitter may be slightly off frequency, your batteries may be running low, and/or you could be in a bad location.*

This is yet another "all of the above" question on the exam. Three things that can cause distorted or unintelligible signals to be received at a repeater (or any receiver) is insufficient power from low batteries, your transmitter may not be transmitting at the desired frequency and needs to be adjusted, or you could just be in a bad spot.

T7B11 What is a symptom of RF feedback in a transmitter or transceiver?
Answer: *Reports of garbled, distorted, or unintelligible transmissions.*

RF feedback can mix with the signal you are trying to send and cause distortion. Your own RF signals can affect your own transmitter just like they can to a television or stereo. This RF feedback does not mix nicely with the signal you are transmitting and can produce garbled, distorted, or unintelligible signals.

T7B12 What does the acronym "BER" mean when applied to digital communications systems?
Answer: *Bit Error Rate.*

Data transmissions, just like voice transmissions, are sometimes hard to hear and unintelligible. When computers have a hard time understanding each other the result is errors in the received data. Bit error rate (BER) is a measure of errors received in a data transmission.

Quiz #24

(Find the answers on page 256)

1. _____ T7B01 What can you do if you are told your FM handheld or mobile transceiver is over deviating?
A. Talk louder into the microphone
B. Let the transceiver cool off
C. Change to a higher power level
D. Talk farther away from the microphone

2. _____ T7B02 What is meant by fundamental overload in reference to a receiver?
A. Too much voltage from the power supply
B. Too much current from the power supply
C. Interference caused by very strong signals
D. Interference caused by turning the volume up too high

3. _____ T7B03 Which of the following may be a cause of radio frequency interference?
A. Fundamental overload
B. Harmonics
C. Spurious emissions
D. All of these choices are correct

4. _____ T7B04 What is the most likely cause of interference to a non-cordless telephone from a nearby transmitter?
A. Harmonics from the transmitter
B. The telephone is inadvertently acting as a radio receiver
C. Poor station grounding
D. Improper transmitter adjustment

5. _____ T7B05 What is a logical first step when attempting to cure a radio frequency interference problem in a nearby telephone?
A. Install a low-pass filter at the transmitter
B. Install a high-pass filter at the transmitter
C. Install an RF filter at the telephone
D. Improve station grounding

6. _____ T7B06 What should you do first if someone tells you that your station's transmissions are interfering with their radio or TV reception?
A. Make sure that your station is functioning properly and that it does not cause interference to your own television
B. Immediately turn off your transmitter and contact the nearest FCC office for assistance
C. Tell them that your license gives you the right to transmit and nothing can be done to reduce the interference
D. Continue operating normally because your equipment cannot possibly cause any interference

7. _____ T7B07 Which of the following may be useful in correcting a radio frequency interference problem?
A. Snap-on ferrite chokes
B. Low-pass and high-pass filters
C. Band-reject and band-pass filters
D. All of these choices are correct

8. _____ T7B08 What should you do if a "Part 15" device in your neighbor's home is causing harmful interference to your amateur station?
A. Work with your neighbor to identify the offending device
B. Politely inform your neighbor about the rules that require him to stop using the device if it causes interference
C. Check your station and make sure it meets the standards of good amateur practice
D. All of these choices are correct

9. _____ T7B09 What could be happening if another operator reports a variable high-pitched whine on the audio from your mobile transmitter?
A. Your microphone is picking up noise from an open window
B. You have the volume on your receiver set too high
C. You need to adjust your squelch control
D. Noise on the vehicle's electrical system is being transmitted along with your speech audio

10. _____ T7B10 What might be the problem if you receive a report that your audio signal through the repeater is distorted or unintelligible?
A. Your transmitter may be slightly off frequency
B. Your batteries may be running low
C. You could be in a bad location
D. All of these choices are correct

11. _____ T7B11 What is a symptom of RF feedback in a transmitter or transceiver?
A. Excessive SWR at the antenna connection
B. The transmitter will not stay on the desired frequency
C. Reports of garbled, distorted, or unintelligible transmissions
D. Frequent blowing of power supply fuses

12. _____ T7B12 What does the acronym "BER" mean when applied to digital communications systems?
A. Baud Enhancement Recovery
B. Baud Error Removal
C. Bit Error Rate
D. Bit Exponent Resource

Lesson 24 Study Sheet

✓ BER stands for Bit Error Rate.

Notes:

Lesson 25
The T7C Section
Antenna Measurements and Troubleshooting

T7C01 What is the primary purpose of a dummy load?
Answer: *To prevent the radiation of signals when making tests.*

A dummy load is a piece of test equipment that allows you to test your transmitter at power without producing a signal. A dummy load connects to the RF output of the transmitter in the place of the antenna. The dummy load absorbs the transmitters signal and produces heat instead of a radio signal. Be careful if you are testing your transmitter at high power. The dummy load can get very hot.

T7C02 Which of the following instruments can be used to determine if an antenna is resonant at the desired operating frequency?
Answer: *An antenna analyzer.*

An antenna analyzer analyzes antennas. An antenna analyzer is a small, low power transmitter that will measure the SWR and impedance of an antenna. If the standing wave ratio (SWR) of the antenna is low (close to 1:1), the antenna is resonant at the frequency you are testing.

T7C03 What, in general terms, is standing wave ratio (SWR)?
Answer: *A measure of how well a load is matched to a transmission line.*

Standing wave ratio (SWR) is a measurement of how well the impedance of a transmitter is matched to the impedance of an antenna and feedline. The ideal SWR is 1:1. If the impedance is poorly matched, the antenna and feedline will reflect some of the energy back to the transmitter. When the energy from the transmitter collides with the power reflected back from the antenna, it causes a standing wave to occur in the feedline. If the power reflected back is large enough it can damage your transmitter.

T7C04 What reading on an SWR meter indicates a perfect impedance match between the antenna and the feedline?
Answer: *1 to 1.*

A 1:1 SWR is a perfect match between the impedance of your transmitter and your antenna/feedline. One of the things this tells you is the antenna is resonant at the transmit frequency and the majority of your transmitting power is generating a radio signal vice radiating out of your feedline.

T7C05 What is the approximate SWR value above which the protection circuits in most solid-state transmitters begin to reduce transmitter power?
Answer: *2 to 1.*

High SWR can damage your transmitter. Most modern transmitters have safety devises built in that will automatically reduce the transmitter's output power once the SWR goes above 2:1. Reducing the output power lowers the amount of energy reflected back to the transmitter and helps prevent damage to the radio's circuits.

T7C06 What does an SWR reading of 4:1 mean?
Answer: *An impedance mismatch.*

Technically, any SWR greater than 1:1 is an impedance mismatch, but 4:1 definitely is. This does not necessarily mean that you are not producing a signal, but that your signal is not efficient for the amount of power the transmitter is using.

T7C07 What happens to power lost in a feedline?
Answer: *It is converted into heat.*

All feedlines have imperfections which cause resistance to current flow. This resistance causes some of the power the transmitter sends to the antenna to be converted to heat.

T7C08 What instrument other than an SWR meter could you use to determine if a feedline and antenna are properly matched?
Answer: *Directional wattmeter.*

A directional wattmeter measures current flowing in a particular direction. The directional wattmeter can measure the power flowing out of the transmitter to the antenna and then reversed to measure the power flowing back to the transmitter from the antenna. These two power measurements will give you the SWR.

T7C09 Which of the following is the most common cause for failure of coaxial cables?
Answer: *Moisture contamination.*

Moisture is the natural enemy of coaxial cable. If the interior of a coaxial cable gets wet, it is ruined. If the outer insulation of coax is damaged or a connection is not properly sealed and water proofed, water can get in and increase resistance or corrode the interior shielding. Take care to protect your coaxial cable from the elements!

T7C10 Why should the outer jacket of coaxial cable be resistant to ultraviolet light?
Answer: *Ultraviolet light can damage the jacket and allow water to enter the cable.*

Ultraviolet light will deteriorate the outer insulation of coaxial cable. There are types of coax you can buy which are resistant to ultraviolet light, but nothing is completely immune. Ultraviolet rays will eventually cause the outer jacket of coax to become brittle and crack. Once it is cracked, moisture can get in. Once the interior of coaxial cable gets wet, it needs to be replaced.

T7C11 What is a disadvantage of "air core" coaxial cable when compared to foam or solid dielectric types?
Answer: *It requires special techniques to prevent water absorption.*

"Air core" coaxial cable has the advantage of exhibiting lower loss than the foam or solid dielectric types of coax. However, it is much higher maintenance. Special care must be made to prevent water absorption at the connectors of "air core" coaxial cable. The advantage is a more efficient signal. The disadvantage is "air core" coax is more susceptible to moisture damage.

Quiz #25

(Find the answers on page 256)

1. _____ T7C01 What is the primary purpose of a dummy load?
A. To prevent the radiation of signals when making tests
B. To prevent over-modulation of your transmitter
C. To improve the radiation from your antenna
D. To improve the signal to noise ratio of your receiver

2. _____ T7C02 Which of the following instruments can be used to determine if an antenna is resonant at the desired operating frequency?
A. A VTVM
B. An antenna analyzer
C. A "Q" meter
D. A frequency counter

3. _____ T7C03 What, in general terms, is standing wave ratio (SWR)?
A. A measure of how well a load is matched to a transmission line
B. The ratio of high to low impedance in a feedline
C. The transmitter efficiency ratio
D. An indication of the quality of your station's ground connection

4. _____ T7C04 What reading on an SWR meter indicates a perfect impedance match between the antenna and the feedline?
A. 2 to 1
B. 1 to 3
C. 1 to 1
D. 10 to 1

5. _____ T7C05 What is the approximate SWR value above which the protection circuits in most solid-state transmitters begin to reduce transmitter power?
A. 2 to 1
B. 1 to 2
C. 6 to 1
D. 10 to 1

6. _____ T7C06 What does an SWR reading of 4:1 mean?
A. An antenna loss of 4 dB
B. A good impedance match
C. An antenna gain of 4
D. An impedance mismatch

7. _____ T7C07 What happens to power lost in a feedline?
A. It increases the SWR
B. It comes back into your transmitter and could cause damage
C. It is converted into heat
D. It can cause distortion of your signal

8. _____ T7C08 What instrument other than an SWR meter could you use to determine if a feedline and antenna are properly matched?
A. Voltmeter
B. Ohmmeter
C. Iambic pentameter
D. Directional wattmeter

9. _____ T7C09 Which of the following is the most common cause for failure of coaxial cables?
A. Moisture contamination
B. Gamma rays
C. The velocity factor exceeds 1.0
D. Overloading

10. _____ T7C10 Why should the outer jacket of coaxial cable be resistant to ultraviolet light?
A. Ultraviolet resistant jackets prevent harmonic radiation
B. Ultraviolet light can increase losses in the cable's jacket
C. Ultraviolet and RF signals can mix together, causing interference
D. Ultraviolet light can damage the jacket and allow water to enter the cable

11. _____ T7C11 What is a disadvantage of "air core" coaxial cable when compared to foam or solid dielectric types?
A. It has more loss per foot
B. It cannot be used for VHF or UHF antennas
C. It requires special techniques to prevent water absorption
D. It cannot be used at below freezing temperatures

Lesson 25 Study Sheet

✓ A dummy load prevents the radiation of signals when testing a transmitter.

✓ An antenna analyzer is used to determine if an antenna is resonant at the desired operating frequency.

✓ Standing wave ratio (SWR) is a measure of how well a load impedance is matched to transmission line impedance. A perfect match has an SWR of 1:1.

Notes:

Lesson 26
The T7D Section
Basic Repair and Testing

T7D01 Which instrument would you use to measure electric potential or electromotive force?
Answer: *A voltmeter.*

Electric potential and electromotive force (EMF) are also known as voltage. To measure voltage you need a voltmeter! The basic unit of measurement of voltage is the volt.

T7D02 What is the correct way to connect a voltmeter to a circuit?
Answer: *In parallel with the circuit.*

Voltage is measured in parallel with the circuit. To accurately measure voltage, you place the leads of the voltmeter across the circuit being measured. Think of this as placing one lead at the beginning of a circuit and the other at the end of the circuit. This will give you an accurate measurement of the circuit's voltage.

T7D03 How is an ammeter usually connected to a circuit?
Answer: *In series with the circuit.*

Ammeters measure electric current. To measure current you place the leads of the ammeter in series with the circuit. To accurately measure current flow, place the leads along the flow of current within the circuit.

T7D04 Which instrument is used to measure electric current?
Answer: *An ammeter.*

An ammeter measures electric current in a circuit. The basic unit of measurement for electric current is the ampere.

T7D05 What instrument is used to measure resistance?
Answer: *An ohmmeter.*

Ohmmeters measure resistance. The basic unit of measurement of resistance is the ohm.

T7D06 Which of the following might damage a multimeter?
Answer: *Attempting to measure voltage when using the resistance setting.*

A multimeter is several different types of meters in one device. A switch on the multimeter usually selects the type of meter to be used. A voltmeter measures the voltage of a circuit while the circuit is powered. An ohmmeter measures resistance while the power to the circuit is off. The ohmmeter generates a small amount of power to measure resistance. If a multimeter is set to measure resistance and is connected to a circuit with power running through it, the result may be a broken multimeter.

T7D07 Which of the following measurements are commonly made using a multimeter?
Answer: *Voltage and resistance.*

Most multimeters will have at least a voltmeter and ohmmeter function.

T7D08 Which of the following types of solder is best for radio and electronic use?
Answer: *Rosin-core solder.*

Rosin-core solder melts easily and is good for more the precise soldering that is needed in electronics. The rosin in the core melts before the solder and flows onto the joint to be soldered. The rosin removes some of the corrosion at the joint prior to the solder melting and helps make a good electrical connection.

T7D09 What is the characteristic appearance of a "cold" solder joint?
Answer: *A grainy or dull surface.*

A grainy or dull surface on a solder joint may be a poor joint. It usually indicates that joint wasn't heated sufficiently for the solder to melt properly and create a good electrical connection.

T7D10 What is probably happening when an ohmmeter, connected across a circuit, initially indicates a low resistance and then shows increasing resistance with time?
Answer: *The circuit contains a large capacitor.*

To measure resistance, an ohmmeter adds energy to the circuit. Remember a capacitor stores energy in an electric field. If a capacitor is part of the circuit the ohmmeter is measuring, the energy the ohmmeter adds to the circuit may build up the capacitor's electric field. As the capacitor's electric field builds up, the resistance the ohmmeter is measuring will increase.

T7D11 Which of the following precautions should be taken when measuring circuit resistance with an ohmmeter?
Answer: *Ensure that the circuit is not powered.*

The circuit must not be powered when measuring its resistance with an ohmmeter. If the circuit is powered when the ohmmeter is attached the ohmmeter may be damaged.

Quiz #26

(Find the answers on page 256)

1. _____ T7D01 Which instrument would you use to measure electric potential or electromotive force?
A. An ammeter
B. A voltmeter
C. A wavemeter
D. An ohmmeter

2. _____ T7D02 What is the correct way to connect a voltmeter to a circuit?
A. In series with the circuit
B. In parallel with the circuit
C. In quadrature with the circuit
D. In phase with the circuit

3. _____ T7D03 How is an ammeter usually connected to a circuit?
A. In series with the circuit
B. In parallel with the circuit
C. In quadrature with the circuit
D. In phase with the circuit

4. _____ T7D04 Which instrument is used to measure electric current?
A. An ohmmeter
B. A wavemeter
C. A voltmeter
D. An ammeter

5. _____ T7D05 What instrument is used to measure resistance?
A. An oscilloscope
B. A spectrum analyzer
C. A noise bridge
D. An ohmmeter

6. _____ T7D06 Which of the following might damage a multimeter?
A. Measuring a voltage too small for the chosen scale
B. Leaving the meter in the milliamps position overnight
C. Attempting to measure voltage when using the resistance setting
D. Not allowing it to warm up properly

7. _____ T7D07 Which of the following measurements are commonly made using a multimeter?
A. SWR and RF power
B. Signal strength and noise
C. Impedance and reactance
D. Voltage and resistance

8. _____ T7D08 Which of the following types of solder is best for radio and electronic use?
A. Acid-core solder
B. Silver solder
C. Rosin-core solder
D. Aluminum solder

9. _____ T7D09 What is the characteristic appearance of a "cold" solder joint?
A. Dark black spots
B. A bright or shiny surface
C. A grainy or dull surface
D. A greenish tint

10. _____ T7D10 What is probably happening when an ohmmeter, connected across a circuit, initially indicates a low resistance and then shows increasing resistance with time?
A. The ohmmeter is defective
B. The circuit contains a large capacitor
C. The circuit contains a large inductor
D. The circuit is a relaxation oscillator

11. _____ T7D11 Which of the following precautions should be taken when measuring circuit resistance with an ohmmeter?
A. Ensure that the applied voltages are correct
B. Ensure that the circuit is not powered
C. Ensure that the circuit is grounded
D. Ensure that the circuit is operating at the correct frequency

Lesson 26 Study Sheet

✓ A voltmeter measures voltage in a circuit. Voltage is also known as electric potential and electromotive force (EMF).

✓ An ammeter measures current in a circuit. The basic unit of measure for current is the ampere.

✓ An ohmmeter measures resistance in a circuit. The basic unit of measure for resistance is the ohm.

✓ Most multimeters will have at least a voltmeter and ohmmeter function.

Notes:

Lesson 27
The T8A Section
Modulation Modes

T8A01 Which of the following is a form of amplitude modulation?
Answer: *Single sideband.*

Single sideband (SSB) is a form of amplitude modulation (AM). AM signals are made of three separate signals: two sideband signals and a carrier signal. One sideband is a higher frequency than the carrier frequency and the other sideband is a lower frequency than the carrier. The higher frequency is called the upper sideband (USB) and the lower frequency sideband is called the lower sideband (LSB). You can send signals using either USB or LSB. As they are one of the two AM sidebands, USB and LSB are categorized as single sidebands (SSB).

T8A02 What type of modulation is most commonly used for VHF packet radio transmissions?
Answer: *FM.*

Frequency modulation (FM) is most commonly used for VHF packet radio. Part of the reason is simply by convention, but FM provides a very clean signal and works very well with data modes.

T8A03 Which type of voice modulation is most often used for long-distance or weak signal contacts on the VHF and UHF bands?
Answer: *SSB.*

Single sideband works very well for long distance voice communications. SSB signals have a much narrower bandwidth than FM signals. The narrower bandwidth gives SSB signals more concentrated power and thus more traveling distance.

T8A04 Which type of modulation is most commonly used for VHF and UHF voice repeaters?
Answer: *FM.*

Frequency modulation (FM) is one of the more popular bands for VHF and UHF voice communication. FM signals are very clean and free of noise compared to other forms of modulation. This makes FM very popular for short distance repeater communication.

T8A05 Which of the following types of emission has the narrowest bandwidth?
Answer: *CW.*

CW (Morse code) communications have a very narrow bandwidth. This allows many CW signals in a relatively small amount of space on the frequency spectrum. This also allows CW signals to travel very long distances with a very small amount of power.

T8A06 Which sideband is normally used for 10 meter HF, VHF and UHF single-sideband communications?
Answer: *Upper sideband.*

Upper sideband (USB) is usually used for the higher frequencies. This is for the most part simply by convention. Lower sideband will work perfectly well, but you may have a hard time finding someone who is listening. The general convention is USB is used for frequencies in the 20 meter band and higher. For frequencies in the 30 meter band and lower, LSB is used.

T8A07 What is the primary advantage of single sideband over FM for voice transmissions?
Answer: *SSB signals have narrower bandwidth.*

SSB signals have a narrower bandwidth than FM. FM signals take up a relatively large amount of space which limits their range and limits the amount of signals that can fit in a frequency range. SSB signals have a narrower bandwidth which allows them to travel farther and keeps the band less crowded with signals.

T8A08 What is the approximate bandwidth of a single sideband voice signal?
Answer: *3 kHz.*

A SSB voice signals takes up about 3 kHz of room on the frequency spectrum. This means that from the center of the signal (the transmitting frequency), the signal can be picked up 1.5 MHz above and below the center frequency. Another SSB signal needs to be a minimum of 3 kHz away to avoid causing interference.

T8A09 What is the approximate bandwidth of a VHF repeater FM phone signal?
Answer: *Between 5 and 15 kHz.*

Compared to an SSB signal, an FM signal is much wider. FM signals can take between 5 and 15 kHz of frequency space.

T8A10 What is the typical bandwidth of analog fast-scan TV transmissions on the 70 cm band?
Answer: *About 6 MHz.*

Analog fast-scan TV can take up to 6 MHz of space. That's 6 megahertz! Analog fast-scan TV takes up a lot of room.

T8A11 What is the approximate maximum bandwidth required to transmit a CW signal?
Answer: *150 Hz.*

A CW signal hardly takes up any space at all! A strong CW signal only takes up 150 Hz! That's hertz! There is a lot of spectrum space for CW enthusiasts!

Quiz #27

(Find the answers on page 256)

1. _____ T8A01 Which of the following is a form of amplitude modulation?
A. Spread-spectrum
B. Packet radio
C. Single sideband
D. Phase shift keying

2. _____ T8A02 What type of modulation is most commonly used for VHF packet radio transmissions?
A. FM
B. SSB
C. AM
D. Spread Spectrum

3. _____ T8A03 Which type of voice modulation is most often used for long-distance or weak signal contacts on the VHF and UHF bands?
A. FM
B. AM
C. SSB
D. PM

4. _____ T8A04 Which type of modulation is most commonly used for VHF and UHF voice repeaters?
A. AM
B. SSB
C. PSK
D. FM

5. _____ T8A05 Which of the following types of emission has the narrowest bandwidth?
A. FM voice
B. SSB voice
C. CW
D. Slow-scan TV

6. _____ T8A06 Which sideband is normally used for 10 meter HF, VHF and UHF single-sideband communications?
A. Upper sideband
B. Lower sideband
C. Suppressed sideband
D. Inverted sideband

7. _____ T8A07 What is the primary advantage of single sideband over FM for voice transmissions?
A. SSB signals are easier to tune
B. SSB signals are less susceptible to interference
C. SSB signals have narrower bandwidth
D. All of these choices are correct

8. _____ T8A08 What is the approximate bandwidth of a single sideband voice signal?
A. 1 kHz
B. 3 kHz
C. 6 kHz
D. 15 kHz

9. _____ T8A09 What is the approximate bandwidth of a VHF repeater FM phone signal?
A. Less than 500 Hz
B. About 150 kHz
C. Between 5 and 15 kHz
D. Between 50 and 125 kHz

10. _____ T8A10 What is the typical bandwidth of analog fast-scan TV transmissions on the 70 cm band?
A. More than 10 MHz
B. About 6 MHz
C. About 3 MHz
D. About 1 MHz

11. _____ T8A11 What is the approximate maximum bandwidth required to transmit a CW signal?
A. 2.4 kHz
B. 150 Hz
C. 1000 Hz
D. 15 kHz

Lesson 27 Study Sheet

✓ Single sideband (SSB) is a form of amplitude modulation (AM).

✓ Upper sideband is the sideband most commonly used on the 10 meter band.

✓ SSB voice signals have a bandwidth of 3 kHz.

✓ FM phone signals have a bandwidth of 5 to 15 kHz.

✓ Analog fast-scan television signals on 70 cm have a bandwidth of 6 MHz.

✓ CW signals have a bandwidth of 150 Hz.

Notes:

Lesson 28
The T8B Section
Amateur Satellite Operation

T8B01 Who may be the control operator of a station communicating through an amateur satellite or space station?
Answer: *Any amateur whose license privileges allow them to transmit on the satellite uplink frequency.*

Satellites work a bit like repeaters. Satellites have an uplink frequency on which they receive signals and a downlink frequency on which they send signals. To operate an amateur satellite, an amateur only needs to have license privileges to operate on the uplink frequency even if the downlink frequency does not fall within the amateur's privileges.

T8B02 [97.313(a)] How much transmitter power should be used on the uplink frequency of an amateur satellite or space station?
Answer: *The minimum amount of power needed to complete the contact.*

The rule for transmitter power is only to use the minimum amount necessary to conduct communications, regardless of whether you are working a satellite or not.

T8B03 Which of the following can be done using an amateur radio satellite?
Answer: *Talk to amateur radio operators in other countries.*

Of the possible choices, the correct one is an amateur may use an amateur satellite to talk to amateurs in other countries. Depending on the satellite, there is a bit more a ham can do. The problem with satellites is that the window of time they are in communication range is often very small. A quick conversation may be all there is time for.

T8B04 Which amateur stations may make contact with an amateur station on the International Space Station using 2 meter and 70 cm band amateur radio frequencies?
Answer: *Any amateur holding a Technician or higher class license.*

For this situation, all amateur license classes that have privileges in the 2 meter and 70 cm bands can talk to the astronauts on the International Space Station. Most of the U.S. astronauts assigned to the International Space Station are hams!

T8B05 What is a satellite beacon?
Answer: *A transmission from a space station that contains information about a satellite.*

A satellite beacon is just like the beacon stations used to gauge propagation conditions, except satellite beacons are moving very fast and are in space. Just like a beacon station, satellite beacons transmit information identifying the satellite. If you can hear the beacon transmission from the satellite, you can probably work the satellite.

T8B06 What can be used to determine the time period during which an amateur satellite or space station can be accessed?
Answer: *A satellite tracking program.*

There are many satellite tracking programs available and with many different features and capabilities. Many are available for free on the internet. Some of the more important things these tracking programs will tell you is when the satellite is in communication range and the satellite's coverage area. This is important information to know as satellites travel at high rates of speed which causes them to be in range for only a short period of time.

T8B07 With regard to satellite communications, what is Doppler shift?
Answer: *An observed change in signal frequency caused by relative motion between the satellite and the earth station.*

Doppler shift in satellite communication is the same phenomenon which explains the increasing pitch of a firetruck's siren as it approaches you on a road. As the firetruck approaches, the sound waves of the siren are slightly compressed causing the siren's pitch to increase. The signal frequencies of satellites work the same way. As the satellite approaches your position, the satellite's signal will increase in frequency, sometimes significantly. As the satellite passes overhead and begins traveling away from you, the frequency will decrease.

T8B08 What is meant by the statement that a satellite is operating in "mode U/V"?
Answer: *The satellite uplink is in the 70 cm band and the downlink is in the 2 meter band.*

Satellites have an uplink receive frequency and a downlink transmit frequency. The "mode U/V" means the uplink frequency is in the UHF range (the "U") and the downlink frequency is in the VHF range (the "V").

T8B09 What causes "spin fading" when referring to satellite signals?
Answer: *Rotation of the satellite and its antennas.*

Some satellite are spinning like tops in their orbit (maybe not exactly like tops). This spinning is done purposely and helps the satellite keep its orientation. As the satellite spins, The satellite's antennas change position which can have an impact on the reception of the satellite on the ground.

T8B10 What do the initials LEO tell you about an amateur satellite?
Answer: *The satellite is in a Low Earth Orbit.*

LEO stands for low Earth orbit. These are satellites whose orbits are relatively close to the Earth. Relatively close means less than 1600 km above the Earth.

T8B11 What is a commonly used method of sending signals to and from a digital satellite?
Answer: *FM Packet.*

FM packet is the most common method of digital communication used with amateur satellites.

Quiz #28

(Find the answers on page 256)

1. _____ T8B01 Who may be the control operator of a station communicating through an amateur satellite or space station?
A. Only an Amateur Extra Class operator
B. A General Class licensee or higher licensee who has a satellite operator certification
C. Only an Amateur Extra Class operator who is also an AMSAT member
D. Any amateur whose license privileges allow them to transmit on the satellite uplink frequency

2. _____ T8B02 [97.313(a)] How much transmitter power should be used on the uplink frequency of an amateur satellite or space station?
A. The maximum power of your transmitter
B. The minimum amount of power needed to complete the contact
C. No more than half the rating of your linear amplifier
D. Never more than 1 watt

3. _____ T8B03 Which of the following can be done using an amateur radio satellite?
A. Talk to amateur radio operators in other countries
B. Get global positioning information
C. Make telephone calls
D. All of these choices are correct

4. _____ T8B04 Which amateur stations may make contact with an amateur station on the International Space Station using 2 meter and 70 cm band amateur radio frequencies?
A. Only members of amateur radio clubs at NASA facilities
B. Any amateur holding a Technician or higher class license
C. Only the astronaut's family members who are hams
D. You cannot talk to the ISS on amateur radio frequencies

5. _____ T8B05 What is a satellite beacon?
A. The primary transmit antenna on the satellite
B. An indicator light that that shows where to point your antenna
C. A reflective surface on the satellite
D. A transmission from a space station that contains information about a satellite

6. _____ T8B06 What can be used to determine the time period during which an amateur satellite or space station can be accessed?
A. A GPS receiver
B. A field strength meter
C. A telescope
D. A satellite tracking program

7. _____ T8B07 With regard to satellite communications, what is Doppler shift?
A. A change in the satellite orbit
B. A mode where the satellite receives signals on one band and transmits on another
C. An observed change in signal frequency caused by relative motion between the satellite and the earth station
D. A special digital communications mode for some satellites

8. _____ T8B08 What is meant by the statement that a satellite is operating in "mode U/V"?
A. The satellite uplink is in the 15 meter band and the downlink is in the 10 meter band
B. The satellite uplink is in the 70 cm band and the downlink is in the 2 meter band
C. The satellite operates using ultraviolet frequencies
D. The satellite frequencies are usually variable

9. _____ T8B09 What causes "spin fading" when referring to satellite signals?
A. Circular polarized noise interference radiated from the sun
B. Rotation of the satellite and its antennas
C. Doppler shift of the received signal
D. Interfering signals within the satellite uplink band

10. _____ T8B10 What do the initials LEO tell you about an amateur satellite?
A. The satellite battery is in Low Energy Operation mode
B. The satellite is performing a Lunar Ejection Orbit maneuver
C. The satellite is in a Low Earth Orbit
D. The satellite uses Light Emitting Optics

11. _____ T8B11 What is a commonly used method of sending signals to and from a digital satellite?
A. USB AFSK
B. PSK31
C. FM Packet
D. WSJT

Lesson 28 Study Sheet

✓ A satellite beacon is a transmission from a space station that contains information about a satellite.

✓ Doppler shift, in regards to satellite communications, is an observed change in signal frequency caused by the relative motion between the satellite and the earth station.

✓ LEO stands for Low Earth Orbit.

Notes:

Lesson 29
The T8C Section
Operating Activities

T8C01 Which of the following methods is used to locate sources of noise interference or jamming?
Answer: *Radio direction finding*.

Locating the source of a signal is called radio direction finding. There are a number of amateur activities associated with radio direction finding, one of the biggest are "fox hunts." "Fox hunts" are hidden transmitter hunts where the goal is to locate a transmitter sending a signal. "Fox hunts" are usually a mix of radio and orienteering skills to increase the challenge of he event.

T8C02 Which of these items would be useful for a hidden transmitter hunt?
Answer: *A directional antenna*.

One of the tools which are necessary for hidden transmitter hunts is a directional antenna. To find the hidden transmitter you need to know what direction to start looking in. A directional antenna focuses on signals in the direction it's pointing. Combine the directional antenna with a signal strength meter and go in the direction that has the strongest signal reading.

T8C03 What popular operating activity involves contacting as many stations as possible during a specified period of time?
Answer: *Contesting*.

Contesting is a popular activity among hams and provides an opportunity to show off your operating skills. Contests usually consist of making as many successful contacts as possible on a specific band with a particular mode. The contact will specify the rules dealing with time limits and the information which needs to be exchanged for each contact. Contesting is a lot of fun and is a good way to improve your operating skills. Furthermore, you can win a lot of plaques and certificates to put on your wall!

T8C04 Which of the following is good procedure when contacting another station in a radio contest?
Answer: *Send only the minimum information needed for proper identification and the contest exchange*.

In a contest, you want to make as many contacts as possible. This means keeping each contact as short as possible. Exchange only the minimum information necessary for the contact.

T8C05 What is a grid locator?
Answer: *A letter-number designator assigned to a geographic location.*

The whole world is divided up into grid squares based on latitude and longitude lines. Each grid square is assigned a series of two letters and two numbers called a grid locator.

T8C06 For what purpose is a temporary "1 by 1" format (letter-number-letter) call sign assigned?
Answer: *For operations in conjunction with an activity of special significance to the amateur community.*

"1 by 1" call signs are temporarily assigned to stations conducting an activity of special significance to the amateur community. These call signs are made of a single letter, followed by a single digit number, and then followed by a single letter. K3A is an example of a special event, "1 by 1" call.

T8C07 [97.215(c)] What is the maximum power allowed when transmitting telecommand signals to radio controlled models?
Answer: *1 watt.*

If you are controlling a radio controlled vehicle on a frequency within amateur privileges, you are not allowed to exceed 1 watt output power from the transmitter.

T8C08 [97.215(a)] What is required in place of on-air station identification when sending signals to a radio control model using amateur frequencies?
Answer: *A label indicating the licensee's name, call sign and address must be affixed to the transmitter.*

When transmitting control commands to a radio controlled vehicle, it is not very practical to stop and send your station call sign. To identify your station call sign, the FCC requires that you have your station information clearly marked on the transmitter used to control the vehicle. This information consists of the licensee's name, call sign, and address.

T8C09 How might you obtain a list of active nodes that use VoIP?
Answer: *From a repeater directory.*

VoIP stands for voice over the internet protocol and is a system which allows voice communication over the internet. You can mix communicating over the internet and amateur radio! One of the ways to do this is through a node. A node is an interface between a transceiver and a computer with access to the internet. These internet and amateur radio interfaces often operate similar to a repeater in that a signal is received on one frequency, sent through the internet, and retransmitted on another frequency. Many repeater directories list these nodes.

T8C10 How do you select a specific IRLP node when using a portable transceiver?
Answer: *Use the keypad to transmit the IRLP node ID.*

IRLP stands for Internet Radio Link Project. To access an IRLP node you need to know the IRLP node's ID. The node's ID is determined by the owner of the node. Once you know the code, you can punch it into the keypad and access the node.

T8C11 What name is given to an amateur radio station that is used to connect other amateur stations to the Internet?
Answer: *A gateway.*

A gateway is an amateur station that provides access to the internet through a transmitter. These are primarily data type transmissions.

Quiz #29

(Find the answers on page 256)

1. _____ T8C01 Which of the following methods is used to locate sources of noise interference or jamming?
A. Echolocation
B. Doppler radar
C. Radio direction finding
D. Phase locking

2. _____ T8C02 Which of these items would be useful for a hidden transmitter hunt?
A. Calibrated SWR meter
B. A directional antenna
C. A calibrated noise bridge
D. All of these choices are correct

3. _____ T8C03 What popular operating activity involves contacting as many stations as possible during a specified period of time?
A. Contesting
B. Net operations
C. Public service events
D. Simulated emergency exercises

4. _____ T8C04 Which of the following is good procedure when contacting another station in a radio contest?
A. Be sure to sign only the last two letters of your call if there is a pileup calling the station
B. Work the station twice to be sure that you are in his log
C. Send only the minimum information needed for proper identification and the contest exchange
D. All of these choices are correct

5. _____ T8C05 What is a grid locator?
A. A letter-number designator assigned to a geographic location
B. A letter-number designator assigned to an azimuth and elevation
C. An instrument for neutralizing a final amplifier
D. An instrument for radio direction finding

6. _____ T8C06 For what purpose is a temporary "1 by 1" format (letter-number-letter) call sign assigned?
A. To designate an experimental station
B. To honor a deceased relative who was a radio amateur
C. For operations in conjunction with an activity of special significance to the amateur community
D. All of these choices are correct

7. _____ T8C07 [97.215(c)] What is the maximum power allowed when transmitting telecommand signals to radio controlled models?
A. 500 milliwatts
B. 1 watt
C. 25 watts
D. 1500 watts

8. _____ T8C08 [97.215(a)] What is required in place of on-air station identification when sending signals to a radio control model using amateur frequencies?
A. Voice identification must be transmitted every 10 minutes
B. Morse code ID must be sent once per hour
C. A label indicating the licensee's name, call sign and address must be affixed to the transmitter
D. A flag must be affixed to the transmitter antenna with the station call sign in 1 inch high letters or larger

9. _____ T8C09 How might you obtain a list of active nodes that use VoIP?
A. From the FCC Rulebook
B. From your local emergency coordinator
C. From a repeater directory
D. From the local repeater frequency coordinator

10. _____ T8C10 How do you select a specific IRLP node when using a portable transceiver?
A. Choose a specific CTCSS tone
B. Choose the correct DSC tone
C. Access the repeater autopatch
D. Use the keypad to transmit the IRLP node ID

11. _____ T8C11 What name is given to an amateur radio station that is used to connect other amateur stations to the Internet?
A. A gateway
B. A repeater
C. A digipeater
D. A beacon

Lesson 29 Study Sheet

✓ A grid locator is a letter-number designator assigned to a geographic location.

✓ The FCC may assign a temporary "1 by 1" format (letter-number-letter) call sign to an activity of special significance to the amateur community.

✓ One watt is the maximum power allowed when transmitting telecommand signals to radio controlled models.

Notes:

Lesson 30
The T8D Section
Non-voice Communications

T8D01 Which of the following is an example of a digital communications method?
Answer: *Packet, PSK31, and MFSK.*

This is an "all of the above" question on the exam. Packet radio, PSK31, and MFSK are all forms of digital communication.

T8D02 What does the term APRS mean?
Answer: *Automatic Position Reporting System.*

APRS stands for Automatic Position Reporting System. APRS is essentially the automatic transmitting of the GPS location information of an amateur station. These stations can be stationary or mobile. This information is usually collected and posted through the internet to provide station locations.

T8D03 Which of the following is normally used when sending automatic location reports via amateur radio?
Answer: *A Global Positioning System receiver.*

To provide reliable location information for a station, the station needs to know where it is. The usual way of doing this is by the transmitter interfacing with a Global Positioning System (GPS) receiver. A GPS calculates its position through a system of satellites. These satellites constantly transmit a signal which the GPS receiver uses to triangulate its position with great precision.

T8D04 What type of transmission is indicated by the term NTSC?
Answer: *An analog fast scan color TV signal.*

To get this answer correct you need to know that NTSC stands for National Television System Committee. The NTSC sets the analog TV standards for many countries to include the United States. There are only two possible answers that deal with television on the exam. The key word you are looking for is "analog." NTSC is associated with analog fast scan TV signals.

T8D05 Which of the following emission modes may be used by a Technician Class operator between 219 and 220 MHz?
Answer: *Data.*

219 to 220 MHz in the 1.25 meter band is set aside specifically for fixed digital message forwarding systems. This is data. This restriction is not just for Technician class operators. This restriction applies to any ham that has 1.25 meter privileges.

T8D06 What does the abbreviation PSK mean?
Answer: *Phase Shift Keying.*

PSK stands for Phase Shift Keying. PSK is a way to send communications by shifting a signal between two frequencies. PSK is a category of data modes characterized by a signal which shifts back and forth between two tones separated by a few Hz: a high tone and low tone. Depending on the specific PSK mode, each letter and number is assigned a specific sequence of high and low tones. The sequence of the high and low tones determines the digital information passed between stations.

T8D07 What is PSK31?
Answer: *A low-rate data transmission mode.*

PSK31 is an increasingly popular data mode. It has a low bandwidth and has the potential for some nice DX data contacts. The problem with PSK31 is it is a bit slow compared to some other digital modes. This means that PSK31 is a low-rate data transmission mode.

T8D08 Which of the following may be included in packet transmissions?
Answer: *A check sum which permits error detection, a header which contains the call sign of the station to which the information is being sent, and automatic repeat request in case of error.*

This is another "all of the above" question on the exam. The reason packet radio is called packet radio is that the digital information is transmitted through a "packet" of information. Each packet has a uniform format which includes a check sum, a header, an automatic repeat request if the message contains errors.

T8D09 What code is used when sending CW in the amateur bands?
Answer: International Morse.

When you think CW, think International Morse code!

T8D10 Which of the following can be used to transmit CW in the amateur bands?
Answer: *A straight Key, an electronic keyer, and a computer Keyboard.*

In the old days of ham radio a straight key (telegraph key) was all there was to send Morse code. The process has become much easier as technology has developed. Now there are electronic keyers which provide a smooth series of "dits and dahs" as well as computer programs which are able to read and send Morse code!

T8D11 What is a "parity" bit?
Answer: *An extra code element used to detect errors in received data.*

A "parity" bit is used to detect errors in data communications. These "parity" bits are extra pieces of code which provide the receiving computer with information to check to see if a particular series of bits were received correctly. It works in a similar manner to a "check" in emergency communications, but instead of a word count the computer is looking for a bit count.

Quiz #30

(Find the answers on page 256)

1. _____ T8D01 Which of the following is an example of a digital communications method?
A. Packet
B. PSK31
C. MFSK
D. All of these choices are correct

2. _____ T8D02 What does the term APRS mean?
A. Automatic Position Reporting System
B. Associated Public Radio Station
C. Auto Planning Radio Set-up
D. Advanced Polar Radio System

3. _____ T8D03 Which of the following is normally used when sending automatic location reports via amateur radio?
A. A connection to the vehicle speedometer
B. A WWV receiver
C. A connection to a broadcast FM sub-carrier receiver
D. A Global Positioning System receiver

4. _____ T8D04 What type of transmission is indicated by the term NTSC?
A. A Normal Transmission mode in Static Circuit
B. A special mode for earth satellite uplink
C. An analog fast scan color TV signal
D. A frame compression scheme for TV signals

5. _____ T8D05 Which of the following emission modes may be used by a Technician Class operator between 219 and 220 MHz?
A. Spread spectrum
B. Data
C. SSB voice
D. Fast-scan television

6. _____ T8D06 What does the abbreviation PSK mean?
A. Pulse Shift Keying
B. Phase Shift Keying
C. Packet Short Keying
D. Phased Slide Keying

7. _____ T8D07 What is PSK31?
A. A high-rate data transmission mode
B. A method of reducing noise interference to FM signals
C. A method of compressing digital television signal
D. A low-rate data transmission mode

8. _____ T8D08 Which of the following may be included in packet transmissions?
A. A check sum which permits error detection
B. A header which contains the call sign of the station to which the information is being sent
C. Automatic repeat request in case of error
D. All of these choices are correct

9. _____ T8D09 What code is used when sending CW in the amateur bands?
A. Baudot
B. Hamming
C. International Morse
D. Gray

10. _____ T8D10 Which of the following can be used to transmit CW in the amateur bands?
A. Straight Key
B. Electronic Keyer
C. Computer Keyboard
D. All of these choices are correct

11. _____ T8D11 What is a "parity" bit?
A. A control code required for automatic position reporting
B. A timing bit used to ensure equal sharing of a frequency
C. An extra code element used to detect errors in received data
D. A "triple width" bit used to signal the end of a character

Lesson 30 Study Sheet

✓ Packet, PSK31, and MFSK are examples of digital communications methods.

✓ APRS stands for Automatic Position Reporting System.

✓ National Television System Committee (NTSC) is associated with analog fast scan color television signals.

✓ PSK stands for Phase Shift Keying.

✓ The only emission mode allowed to licensed amateurs between 219 and 220 MHz is data.

✓ PSK31 is a low-rate data transmission mode.

✓ A "parity" bit is an extra code element used to detect errors in received data.

Notes:

Lesson 31
The T9A Section
Antennas

T9A01 What is a beam antenna?
Answer: *An antenna that concentrates signals in one direction.*

A beam antenna is a type of directional antenna. A beam antenna acts like a lens, focusing the signal from your transmitter in a particular direction. This concentrates the energy of the signal in the desired direction potentially allowing your signal to travel longer distances. Similar to focusing your transmitted signal in a direction, a beam antenna can also focus on signals coming from a particular direction. This allows the receiver to hone in on a desired signal and avoid undesired signals coming from other directions.

T9A02 Which of the following is true regarding vertical antennas?
Answer: *The electric field is perpendicular to the Earth.*

Vertical antennas are positioned straight up and down, or in other words, vertically. The electric field they generate when transmitting is also vertically oriented. Therefore, the electric field is perpendicular to the Earth. Be careful with this question! One of the possible answers on the exam describes the antenna's magnetic field as being perpendicular to the Earth. The correct answer is electric field!

T9A03 Which of the following describes a simple dipole mounted so the conductor is parallel to the Earth's surface?
Answer: *A horizontally polarized antenna.*

If the antenna is mounted so the conductors are parallel to the Earth's surface, it is horizontally mounted. That makes it a horizontally polarized antenna!

T9A04 What is a disadvantage of the "rubber duck" antenna supplied with most handheld radio transceivers?
Answer: *It does not transmit or receive as effectively as a full-sized antenna.*

"Rubber duck" antennas are the small portable antennas that are associated with handheld VHF/UHF transceivers. "Rubber duck" antennas work fine for short range, low power communications, but they are nowhere near as efficient or effective as a full-sized antenna. The general rule with antennas is the longer the better.

T9A05 How would you change a dipole antenna to make it resonant on a higher frequency?
Answer: *Shorten it.*

The length of a dipole antenna is directly proportional to the wavelength of the tuned frequency. What this means is the longer the wavelength, the longer the dipole. As frequency increases the wavelength of the radio signal decreases. To adjust a dipole antenna to resonate at a higher frequency the dipole just needs to be shortened.

T9A06 What type of antennas are the quad, Yagi, and dish?
Answer: *Directional antennas.*

Quads, Yagis, and dish antennas all focus radio signals in a particular direction. This makes them directional antennas. This is an "all of the above" question on the exam.

T9A07 What is a good reason not to use a "rubber duck" antenna inside your car?
Answer: *Signals can be significantly weaker than when it is outside of the vehicle.*

As we talked about earlier, "rubber duck" antennas are nowhere near the best in the world. They are even worse inside a small metal container, like your car. If you are going to use your handheld transceiver in the car, hook it up to an externally mounted mobile antenna.

T9A08 What is the approximate length, in inches, of a quarter-wavelength vertical antenna for 146 MHz?
Answer: *19.*

There is a formula for calculating antenna length based on frequency. The formula used to calculate a half-wave dipole (a dipole equivalent to half the frequency's wavelength) is:

Length (in feet) = 468/frequency (in MHz)

Let's start by calculating the half-wavelength.

Length (in feet) = 468/146 MHz
Length = 3.2 feet

The question is asking for inches. Since there are 12 inches per foot, multiply 3.2 feet by 12 inches per foot.

Length (in inches) = 3.2 feet x 12 inches/foot
Length = 38.4 inches

This is the length of half of the frequency's wavelength in inches. However, the question is asking the length of a quarter-wavelength antenna. The quarter-wavelength of a frequency is half of the half-wavelength of the frequency. To find the quarter-wavelength just divide the half-wavelength by 2.

38.4 inches/2 = 19.2 inches

The closest answer on the exam is 19 inches!

T9A09 What is the approximate length, in inches, of a 6 meter 1/2-wavelength wire dipole antenna?
Answer: *112*.

There are two ways to solve this question. The first is to estimate the half-wavelength in meters and then convert to inches. The second is to pick a 6 meter frequency and calculate using the half-wavelength formula.

You already know the approximate wavelength of a frequency in the 6 meter band. It's 6 meters! The half-wavelength is 3 meters. There are 3.1 feet per meter so multiply 3 by 3.1 which equals 9.3 feet. Multiply 9.3 by 12 inches and you get 111.6 inches. The closest answer is 112 inches.

Now let's use the formula.

Length (in feet) = 468/frequency (in MHz)

The amateur privileges in the 6 meter band range from 50 MHz to 54 MHz. We will pick a frequency in the middle, 53 MHz.

Length (in feet) = 468/52 MHz
Length = 9 feet

The question is asking for inches so multiply by 12 inches.

Length (in inches) = 9 feet x 12 inches/feet
Length = 108 inches

The closest answer is 112 inches!

T9A10 In which direction is the radiation strongest from a half-wave dipole antenna in free space?
Answer: *Broadside to the antenna*.

If a dipole is hung from north to south, the signal radiates strongest to the east and west. The signal radiates strongest in the direction perpendicular to the wire of the dipole antenna.

T9A11 What is meant by the gain of an antenna?
Answer: *The increase in signal strength in a specified direction when compared to a reference antenna.*

You may be able to hear a signal with an omnidirectional antenna, but you may be able to hear it better if you point a directional antenna in the direction of the signal. In this example, the increased signal strength of the directional antenna, as compared to the omnidirectional antenna, is the gain of the directional antenna.

Quiz #31

(Find the answers on page 256)

1. _____ T9A01 What is a beam antenna?
A. An antenna built from aluminum I-beams
B. An omnidirectional antenna invented by Clarence Beam
C. An antenna that concentrates signals in one direction
D. An antenna that reverses the phase of received signals

2. _____ T9A02 Which of the following is true regarding vertical antennas?
A. The magnetic field is perpendicular to the Earth
B. The electric field is perpendicular to the Earth
C. The phase is inverted
D. The phase is reversed

3. _____ T9A03 Which of the following describes a simple dipole mounted so the conductor is parallel to the Earth's surface?
A. A ground wave antenna
B. A horizontally polarized antenna
C. A rhombic antenna
D. A vertically polarized antenna

4. _____ T9A04 What is a disadvantage of the "rubber duck" antenna supplied with most handheld radio transceivers?
A. It does not transmit or receive as effectively as a full-sized antenna
B. It transmits a circularly polarized signal
C. If the rubber end cap is lost it will unravel very quickly
D. All of these choices are correct

5. _____ T9A05 How would you change a dipole antenna to make it resonant on a higher frequency?
A. Lengthen it
B. Insert coils in series with radiating wires
C. Shorten it
D. Add capacity hats to the ends of the radiating wires

6. _____ T9A06 What type of antennas are the quad, Yagi, and dish?
A. Non-resonant antennas
B. Loop antennas
C. Directional antennas
D. Isotropic antennas

7. _____ T9A07 What is a good reason not to use a "rubber duck" antenna inside your car?
A. Signals can be significantly weaker than when it is outside of the vehicle
B. It might cause your radio to overheat
C. The SWR might decrease, decreasing the signal strength
D. All of these choices are correct

8. _____ T9A08 What is the approximate length, in inches, of a quarter-wavelength vertical antenna for 146 MHz?
A. 112
B. 50
C. 19
D. 12

9. _____ T9A09 What is the approximate length, in inches, of a 6 meter 1/2-wavelength wire dipole antenna?
A. 6
B. 50
C. 112
D. 236

10. _____ T9A10 In which direction is the radiation strongest from a half-wave dipole antenna in free space?
A. Equally in all directions
B. Off the ends of the antenna
C. Broadside to the antenna
D. In the direction of the feedline

11. _____ T9A11 What is meant by the gain of an antenna?
A. The additional power that is added to the transmitter power
B. The additional power that is lost in the antenna when transmitting on a higher frequency
C. The increase in signal strength in a specified direction when compared to a reference antenna
D. The increase in impedance on receive or transmit compared to a reference antenna

Lesson 31 Study Sheet

✓ A beam antenna is an antenna that concentrates signals in one direction.

✓ Quad, Yagi, and dish antennas are all types of directional antennas.

✓ To calculate the length of a half-wave dipole use the following formula:

Length (in feet) = 468 / frequency (in MHz)

✓ There are 3.1 feet per meter.

Notes:

Lesson 32
The T9B Section
Feedlines

T9B01 Why is it important to have a low SWR in an antenna system that uses coaxial cable feedline?
Answer: *To allow the efficient transfer of power and reduce losses.*

The higher the standing wave ratio (SWR) the less of your signal is getting out through the antenna. Ensuring a good impedance match between the transmitter and antenna will allow for efficient transfer of power and reduce losses.

T9B02 What is the impedance of the most commonly used coaxial cable in typical amateur radio installations?
Answer: *50 ohms.*

Not all coaxial cable is made the same. Coaxial cables vary in impedance based on what they are used for. For exam purposes, the coaxial cable impedance you need to remember is 50 ohms.

T9B03 Why is coaxial cable used more often than any other feedline for amateur radio antenna systems?
Answer: *It is easy to use and requires few special installation considerations.*

Coaxial cable is easier to use and less maintenance than other types of feedline. There are other types of feedline that work better than coax, but coax is definitely the simplest to use.

T9B04 What does an antenna tuner do?
Answer: *It matches the antenna system impedance to the transceiver's output impedance.*

Antenna tuners are very handy pieces of gear to have around. An antenna tuner will help decrease the SWR between a transmitter and antenna so that they will work well together. An antenna tuner is connected to the transmitter's output and the antenna feedline is connected to the antenna tuner. The antenna tuner can be adjusted until the lowest possible SWR is reached. This "tricks" the transmitter into thinking it's impedance is matched to the antenna and feedline.

T9B05 What generally happens as the frequency of a signal passing through coaxial cable is increased?
Answer: *The loss increases.*

Feedline loss increases as the frequency increases. All types of feedlines experience this to some extent.

T9B06 Which of the following connectors is most suitable for frequencies above 400 MHz?
Answer: *A Type N connector.*

There are different types of connectors for different frequencies. The Type N connector is best for UHF use. There are two types of connectors you need to know for the exam. The Type N connector for UHF work and the PL-259 connector for HF work.

T9B07 Which of the following is true of PL-259 type coax connectors?
Answer: *The are commonly used at HF frequencies.*

PL-259 connectors are primarily used for HF frequencies.

T9B08 Why should coax connectors exposed to the weather be sealed against water intrusion?
Answer: *To prevent an increase in feedline loss.*

Water and coaxial cable do not mix. When moisture invades the interior of coaxial cable, the conductivity of the cable is seriously degraded causing signal losses to drastically increase. Making sure coaxial cable exposed to the elements is properly waterproofed will save you a great deal of heart ache in the long run!

T9B09 What might cause erratic changes in SWR readings?
Answer: *A loose connection in an antenna or a feedline.*

When SWR readings are inconsistent or erratic, check to make sure the antenna and feedline connections are solid. Loose connections can increase feedline impedance and wreak havoc on your signal.

T9B10 What electrical difference exists between the smaller RG-58 and larger RG-8 coaxial cables?
Answer: *RG-8 cable has less loss at a given frequency.*

As was mentioned before, not all coaxial cable is made the same. RG-58 and RG-8 are types of coaxial cable. RG-8 coax has a much wider diameter than RG-58. RG-8's wider diameter results in less loss than the narrower RG-58.

T9B11 Which of the following types of feedline has the lowest loss at VHF and UHF?
Answer: *Air-insulated hard line.*

Air-insulated hard line is hard. It's not the most convenient of feedlines because it is not flexible however, it is the best to use for VHF and UHF in terms of preventing loss.

Quiz #32

(Find the answers on page 256)

1. _____ T9B01 Why is it important to have a low SWR in an antenna system that uses coaxial cable feedline?
A. To reduce television interference
B. To allow the efficient transfer of power and reduce losses
C. To prolong antenna life
D. All of these choices are correct

2. _____ T9B02 What is the impedance of the most commonly used coaxial cable in typical amateur radio installations?
A. 8 ohms
B. 50 ohms
C. 600 ohms
D. 12 ohms

3. _____ T9B03 Why is coaxial cable used more often than any other feedline for amateur radio antenna systems?
A. It is easy to use and requires few special installation considerations
B. It has less loss than any other type of feedline
C. It can handle more power than any other type of feedline
D. It is less expensive than any other types of feedline

4. _____ T9B04 What does an antenna tuner do?
A. It matches the antenna system impedance to the transceiver's output impedance
B. It helps a receiver automatically tune in weak stations
C. It allows an antenna to be used on both transmit and receive
D. It automatically selects the proper antenna for the frequency band being used

5. _____ T9B05 What generally happens as the frequency of a signal passing through coaxial cable is increased?
A. The apparent SWR increases
B. The reflected power increases
C. The characteristic impedance increases
D. The loss increases

6. _____ T9B06 Which of the following connectors is most suitable for frequencies above 400 MHz?
A. A UHF (PL-259/SO-239) connector
B. A Type N connector
C. An RS-213 connector
D. A DB-23 connector

7. _____ T9B07 Which of the following is true of PL-259 type coax connectors?
A. They are good for UHF frequencies
B. They are water tight
C. The are commonly used at HF frequencies
D. They are a bayonet type connector

8. _____ T9B08 Why should coax connectors exposed to the weather be sealed against water intrusion?
A. To prevent an increase in feedline loss
B. To prevent interference to telephones
C. To keep the jacket from becoming loose
D. All of these choices are correct

9. _____ T9B09 What might cause erratic changes in SWR readings?
A. The transmitter is being modulated
B. A loose connection in an antenna or a feedline
C. The transmitter is being over-modulated
D. Interference from other stations is distorting your signal

10. _____ T9B10 What electrical difference exists between the smaller RG-58 and larger RG-8 coaxial cables?
A. There is no significant difference between the two types
B. RG-58 cable has less loss at a given frequency
C. RG-8 cable has less loss at a given frequency
D. RG-58 cable can handle higher power levels

11. _____ T9B11 Which of the following types of feedline has the lowest loss at VHF and UHF?
A. 50-ohm flexible coax
B. Multi-conductor unbalanced cable
C. Air-insulated hard line
D. 75-ohm flexible coax

Lesson 32 Study Sheet

✓ 50 ohms is the impedance of the most commonly used coaxial cable in typical amateur radio installations.

✓ An antenna tuner matches the antenna system impedance to the transceiver's output impedance.

✓ Feedline loss increases as the frequency increases.

✓ A Type N connector is most suitable for frequencies above 400 MHz.

✓ PL-259 coaxial connectors are commonly used for HF frequencies.

✓ RG-58 and RG-8 are types of coaxial cable. Compared to RG-58, RG-8 has less loss at a given frequency.

✓ Air-insulated hard line has the lowest loss when used as feedline for VHF and UHF use.

Notes:

Lesson 33
The T0A Section
AC Power Circuits

T0A01 Which is a commonly accepted value for the lowest voltage that can cause a dangerous electric shock?
Answer: *30 volts.*

Unlike hobbies such as stamp collecting or butterfly catching, amateur radio can hurt or kill you if you are not careful. One of the biggest risks of injury in ham radio is from electric shock. The accepted voltage that is considered dangerous is 30 volts. Remember, get hit with thirty and they'll bury you in the dirt-EE! 30 volts is not much and most of your equipment will probably use more, so be careful!

T0A02 How does current flowing through the body cause a health hazard?
Answer: *By heating tissue, it disrupts the electrical functions of cells, and it causes involuntary muscle contractions.*

Nothing good happens to your body when you are electrocuted. Excessive current running through the body causes tissue to heat up, disrupts the normal electrical function of cells, and causes involuntary muscle contractions. This is an "all of the above" question on the exam.

T0A03 What is connected to the green wire in a three-wire electrical AC plug?
Answer: *Safety ground.*

This is the accepted standard. The ground wire in a three-wire electrical AC plug is green. Sometimes they are just bare wire, but most often the ground wire is green.

T0A04 What is the purpose of a fuse in an electrical circuit?
Answer: *To interrupt power in case of overload.*

A fuse contains a thin piece of metal designed to burn out or melt when the current running through it exceeds a predetermined limit. When that piece of metal burns out, the circuit current stops flowing. Fuses are installed in electronic devices and home electrical systems to prevent excess current from damaging the electronic devices or prevent fires from overheated home wiring.

T0A05 Why is it unwise to install a 20-ampere fuse in the place of a 5-ampere fuse?
Answer: *Excessive current could cause a fire.*

A 5 ampere fuse is designed to melt at or about 5 amperes. The reasoning behind using the 5 ampere fuse for a particular circuit is current in excess of 5 amperes may seriously damage the circuit or start a fire. Replacing a 5 ampere fuse with a 20 ampere fuse may allow a dangerous amount of current to flow through the circuit. This excessive current flow may overheat the circuit and start a fire!

T0A06 What is a good way to guard against electrical shock at your station?
Answer: *Use three-wire cords and plugs for all AC powered equipment, connect all AC powered station equipment to a common safety ground, and use a circuit protected by a ground-fault interrupter.*

In other words, make sure everything is grounded. There are many reasons for having a good ground for radio equipment. Probably the most important of which is to protect against electrical shock. This is an "all of the above" question on the exam.

T0A07 Which of these precautions should be taken when installing devices for lightning protection in a coaxial cable feedline?
Answer: *Ground all of the protectors to a common plate which is in turn connected to an external ground.*

Lightning is the amateur operator's nemesis. Taking precautions to protect your home and equipment against lightning strikes must be your first priority. Coax feedlines should be grounded to a common plate. Ideally you want all your coax entering your home in one location and all the coax should use a single grounding plate. This grounding plate needs to be connected to a grounding rod by the shortest and straightest path possible.

T0A08 What is one way to recharge a 12-volt lead-acid station battery if the commercial power is out?
Answer: *Connect the battery to a car's battery and run the engine.*

A car recharges its battery whenever the engine is running. A car battery is a type of 12-volt lead-acid battery. If the commercial power is out and you have a 12-volt lead-acid battery that needs charging, a running vehicle is an excellent way of recharging it.

T0A09 What kind of hazard is presented by a conventional 12-volt storage battery?
Answer: *Explosive gas can collect if not properly vented.*

12-volt lead-acid batteries produce hydrogen gas (among other things) when they recharge. Recharging a 12-volt storage battery in an unventilated area can result in a dangerous build up of hydrogen which can ignite and explode.

T0A10 What can happen if a lead-acid storage battery is charged or discharged too quickly?
Answer: *The battery could overheat and give off flammable gas or explode.*

Failure to follow proper procedure when recharging a lead-acid battery can cause the battery to overheat and explode. A tool or piece of metal that falls across the battery's positive and negative leads causing the battery to quickly discharge can end with the same results.

T0A11 Which of the following is good practice when installing ground wires on a tower for lightning protection?
Answer: *Ensure that connections are short and direct.*

When installing ground wires, the connection to the ground should be as short and as straight as possible. The goal is to provide the lightning the path of least resistance to the ground. Long ground wires and sharp turns should be avoided to help accommodate the easiest path away from your home and equipment.

T0A12 What kind of hazard might exist in a power supply when it is turned off and disconnected?
Answer: *You might receive an electric shock from stored charge in large capacitors.*

Capacitors store energy in an electric field. When the power to the capacitor is turned off the charge in the capacitor's electric field doesn't disappear. The charge is released back into the circuit over a short period of time. This usually happens fairly quickly, but the larger the capacitor, the longer the charge will take to dissipate. Power supplies may contain large capacitors. It may be quite shocking if you start tinkering with a power supply before allowing enough time for the capacitors to discharge!

T0A13 What safety equipment should always be included in home-built equipment that is powered from 120V AC power circuits?
Answer: *A fuse or circuit breaker in series with the AC "hot" conductor.*

Commercial electrical equipment is built to specific safety standards. Home-built equipment is not required to meet the same standards. When home-built equipment is connected to 120 volt AC certain steps need to be taken to ensure your safety and the safety of those around you. Installing a proper fuse or circuit breaker will help ensure that a safe amount of current flows through the equipment and will stop the current if things start to get unsafe. The important words in the question are “safety equipment.” Of the possible questions, the only answer that makes any mention of items used for safety is the correct one.

Quiz #33

(Find the answers on page 256)

1. _____ T0A01 Which is a commonly accepted value for the lowest voltage that can cause a dangerous electric shock?
A. 12 volts
B. 30 volts
C. 120 volts
D. 300 volts

2. _____ T0A02 How does current flowing through the body cause a health hazard?
A. By heating tissue
B. It disrupts the electrical functions of cells
C. It causes involuntary muscle contractions
D. All of these choices are correct

3. _____ T0A03 What is connected to the green wire in a three-wire electrical AC plug?
A. Neutral
B. Hot
C. Safety ground
D. The white wire

4. _____ T0A04 What is the purpose of a fuse in an electrical circuit?
A. To prevent power supply ripple from damaging a circuit
B. To interrupt power in case of overload
C. To limit current to prevent shocks
D. All of these choices are correct

5. _____ T0A05 Why is it unwise to install a 20-ampere fuse in the place of a 5-ampere fuse?
A. The larger fuse would be likely to blow because it is rated for higher current
B. The power supply ripple would greatly increase
C. Excessive current could cause a fire
D. All of these choices are correct

6. _____ T0A06 What is a good way to guard against electrical shock at your station?
A. Use three-wire cords and plugs for all AC powered equipment
B. Connect all AC powered station equipment to a common safety ground
C. Use a circuit protected by a ground-fault interrupter
D. All of these choices are correct

7. _____ T0A07 Which of these precautions should be taken when installing devices for lightning protection in a coaxial cable feedline?
A. Include a parallel bypass switch for each protector so that it can be switched out of the circuit when running high power
B. Include a series switch in the ground line of each protector to prevent RF overload from inadvertently damaging the protector
C. Keep the ground wires from each protector separate and connected to station ground
D. Ground all of the protectors to a common plate which is in turn connected to an external ground

8. _____ T0A08 What is one way to recharge a 12-volt lead-acid station battery if the commercial power is out?
A. Cool the battery in ice for several hours
B. Add acid to the battery
C. Connect the battery to a car's battery and run the engine
D. All of these choices are correct

9. _____ T0A09 What kind of hazard is presented by a conventional 12-volt storage battery?
A. It emits ozone which can be harmful to the atmosphere
B. Shock hazard due to high voltage
C. Explosive gas can collect if not properly vented
D. All of these choices are correct

10. _____ T0A10 What can happen if a lead-acid storage battery is charged or discharged too quickly?
A. The battery could overheat and give off flammable gas or explode
B. The voltage can become reversed
C. The “memory effect” will reduce the capacity of the battery
D. All of these choices are correct

11. _____ T0A11 Which of the following is good practice when installing ground wires on a tower for lightning protection?
A. Put a loop in the ground connection to prevent water damage to the ground system
B. Make sure that all bends in the ground wires are clean, right angle bends
C. Ensure that connections are short and direct
D. All of these choices are correct

12. _____ T0A12 What kind of hazard might exist in a power supply when it is turned off and disconnected?
A. Static electricity could damage the grounding system
B. Circulating currents inside the transformer might cause damage
C. The fuse might blow if you remove the cover
D. You might receive an electric shock from stored charge in large capacitors

13. _____ T0A13 What safety equipment should always be included in home-built equipment that is powered from 120V AC power circuits?
A. A fuse or circuit breaker in series with the AC "hot" conductor
B. An AC voltmeter across the incoming power source
C. An inductor in series with the AC power source
D. A capacitor across the AC power source

Lesson 33 Study Sheet

✓ 30 volts is the commonly accepted value for the lowest voltage that can cause a dangerous electric shock.

✓ The safety ground is connected to the green wire in a three-wire electrical AC plug.

Notes:

Lesson 34
The T0B Section
Antenna Installation

T0B01 When should members of a tower work team wear a hard hat and safety glasses?
Answer: *At all times when any work is being done on the tower.*

When dealing with matters of safety, it is usually safe to go with the most uncompromising answer on the exam. When working on a tower work team items like tools, branches, and pieces of antenna often come falling down. When working with antennas it is always important to wear safety glasses. Eyes poked by tree branches and antenna elements are one of the most common amateur radio associated injuries.

T0B02 What is a good precaution to observe before climbing an antenna tower?
Answer: *Put on a climbing harness and safety glasses.*

When climbing a tower, your biggest fear should be falling. A climbing harness provides a little peace of mind. Don't forget the safety glasses!

T0B03 Under what circumstances is it safe to climb a tower without a helper or observer?
Answer: *Never.*

Always work on a tower as part of a team! Tower injuries can leave you very unconscious or immobile which is the result of falling from heights. It is also good to have a spotter to help see potential risks before they happen to you.

T0B04 Which of the following is an important safety precaution to observe when putting up an antenna tower?
Answer: *Look for and stay clear of any overhead electrical wires.*

Stay very clear of overhead power lines! You need to make sure the location chosen for the tower will not ever accidentally come in contact with power lines. This includes the lines going to and from your home. Accidental contact with a power line can kill you!

T0B05 What is the purpose of a gin pole?
Answer: *To lift tower sections or antennas.*

A gin pole helps lift tower sections and antennas. A gin pole can be thought of as a mix of a small crane and temporary support for securing tower sections. A gin pole attaches to the top section of an antenna tower. The gin pole extends well above the top section allowing for the next tower section to be raised and put in place. The gin pole is then moved to the top of the new section. That process is repeated until the tower is complete.

T0B06 What is the minimum safe distance from a power line to allow when installing an antenna?
Answer: *So that if the antenna falls unexpectedly, no part of it can come closer than 10 feet to the power wires.*

Sometimes antennas fall. When deciding on a location for your antenna you need to make sure that if the antenna falls, no part of the feedlines or antenna can fall within 10 feet of any power lines.

T0B07 Which of the following is an important safety rule to remember when using a crank-up tower?
Answer: *This type of tower must never be climbed unless it is in the fully retracted position.*

Crank-up towers are designed to provide ease of access. There should be no need to climb a crank-up tower as you can lower the tower to get access to the top section. Crank-up towers are not as sturdy as traditional towers and excess weight may cause them to crank themselves down, sometimes quickly. Even a small shift down can result in a hand or foot getting caught in the tower and causing injury.

T0B08 What is considered to be a proper grounding method for a tower?
Answer: *Separate eight-foot long ground rods for each tower leg, bonded to the tower and each other.*

The goal for safety grounding is to provide the lightning a quick and easy path to the ground. The more grounding rods to disperse the charge into the ground, the better. An eight foot grounding rod for every section of tower with each grounding rod bonded together will help dissipate the lightning and avoid the lightning traveling through the feedline to the equipment.

T0B09 Why should you avoid attaching an antenna to a utility pole?
Answer: *The antenna could contact high-voltage power wires.*

The utility company's utility poles are often used to support power and telephone lines. The golden rule of antennas is to keep them away from power lines.

T0B10 Which of the following is true concerning grounding conductors used for lightning protection?
Answer: *Sharp bends must be avoided.*

Avoid sharp bends when installing grounding conductors for lightning protection. The goal is to have the conductors installed in the shortest route possible allowing the lighting the straightest and least resistant path on its trip to the grounding rods.

T0B11 Which of the following establishes grounding requirements for an amateur radio tower or antenna?
Answer: *Local electrical codes.*

Local electric codes determine grounding requirements for amateur radio towers and antennas. Often times you can find the local electrical codes by contacting your city or county government offices, local fire departments, or even the local hardware store. You can also find some of your ham friends who have a tower and ask them.

Quiz #34

(Find the answers on page 256)

1. _____ T0B01 When should members of a tower work team wear a hard hat and safety glasses?
A. At all times except when climbing the tower
B. At all times except when belted firmly to the tower
C. At all times when any work is being done on the tower
D. Only when the tower exceeds 30 feet in height

2. _____ T0B02 What is a good precaution to observe before climbing an antenna tower?
A. Make sure that you wear a grounded wrist strap
B. Remove all tower grounding connections
C. Put on a climbing harness and safety glasses
D. All of the these choices are correct

3. _____ T0B03 Under what circumstances is it safe to climb a tower without a helper or observer?
A. When no electrical work is being performed
B. When no mechanical work is being performed
C. When the work being done is not more than 20 feet above the ground
D. Never

4. _____ T0B04 Which of the following is an important safety precaution to observe when putting up an antenna tower?
A. Wear a ground strap connected to your wrist at all times
B. Insulate the base of the tower to avoid lightning strikes
C. Look for and stay clear of any overhead electrical wires
D. All of these choices are correct

5. _____ T0B05 What is the purpose of a gin pole?
A. To temporarily replace guy wires
B. To be used in place of a safety harness
C. To lift tower sections or antennas
D. To provide a temporary ground

6. _____ T0B06 What is the minimum safe distance from a power line to allow when installing an antenna?
A. Half the width of your property
B. The height of the power line above ground
C. 1/2 wavelength at the operating frequency
D. So that if the antenna falls unexpectedly, no part of it can come closer than 10 feet to the power wires

7. _____ T0B07 Which of the following is an important safety rule to remember when using a crank-up tower?
A. This type of tower must never be painted
B. This type of tower must never be grounded
C. This type of tower must never be climbed unless it is in the fully retracted position
D. All of these choices are correct

8. _____ T0B08 What is considered to be a proper grounding method for a tower?
A. A single four-foot ground rod, driven into the ground no more than 12 inches from the base
B. A ferrite-core RF choke connected between the tower and ground
C. Separate eight-foot long ground rods for each tower leg, bonded to the tower and each other
D. A connection between the tower base and a cold water pipe

9. _____ T0B09 Why should you avoid attaching an antenna to a utility pole?
A. The antenna will not work properly because of induced voltages
B. The utility company will charge you an extra monthly fee
C. The antenna could contact high-voltage power wires
D. All of these choices are correct

10. _____ T0B10 Which of the following is true concerning grounding conductors used for lightning protection?
A. Only non-insulated wire must be used
B. Wires must be carefully routed with precise right-angle bends
C. Sharp bends must be avoided
D. Common grounds must be avoided

11. _____ T0B11 Which of the following establishes grounding requirements for an amateur radio tower or antenna?
A. FCC Part 97 Rules
B. Local electrical codes
C. FAA tower lighting regulations
D. Underwriters Laboratories' recommended practices

Lesson 34 Study Sheet

✓ A gin pole is used to lift tower sections or antennas.

✓ Separate eight-foot long ground rods for each tower leg, bonded to the tower and each other is considered to be a proper grounding method for a tower.

Notes:

Lesson 35
The T0C Section
RF Hazards

T0C01 What type of radiation are VHF and UHF radio signals?
Answer: *Non-ionizing radiation.*

Non-ionizing radiation is radiation that does not have the capability to ionize atoms. Radiation that is capable of ionizing atoms have a much smaller wavelength than anything in the amateur radio spectrum. Ultraviolet light and x-rays are types of radiation capable causing an atom to loose and electron and become an ion. VHF and UHF signals are non-ionizing forms of radiation. This does not mean there are no risks to prolonged exposure to VHF and UHF signals.

T0C02 Which of the following frequencies has the lowest Maximum Permissible Exposure limit?
Answer: *50 MHz.*

The Maximum Permissible Exposure (MPE) limit defines how much RF radiation a person can be exposed to without causing harm. Like an antenna, the human body resonates at certain frequencies. Generally, people resonate somewhere between 35 MHz and 70 MHz. The human body absorbs more RF energy between these two frequencies than other frequencies. Exposure to RF between these frequencies should be watched closely. 50 MHz is the only possible answer on the exam that falls between these two frequencies.

T0C03 What is the maximum power level that an amateur radio station may use at VHF frequencies before an RF exposure evaluation is required?
Answer: *50 watts PEP at the antenna.*

PEP stands for Peak Envelope Power and is the average of how much power a transmitter sends to an antenna. Essentially, PEP is how many watts you are transmitting. Up to 50 watts on VHF is deemed fairly safe in terms of exposure. Operating above 50 watts requires an RF exposure validation to ensure there is no potential harm to you or those around you from the increased RF radiation.

T0C04 What factors affect the RF exposure of people near an amateur station antenna?
Answer: *Frequency and power level of the RF field, distance from the antenna to a person, and radiation pattern of the antenna.*

This is one of those "all of the above" answers. There are three things that need to be considered when determining RF exposure. The human body will resonate and absorb more radiation at certain frequencies, therefore frequency and power of the RF need to be considered. The proximity to the antenna is important to consider because the RF field is the most concentrated around the antenna. The radiation pattern also needs to be considered. A directional antenna pointed at your neighbors living room may expose them to unsafe levels of RF radiation.

T0C05 Why do exposure limits vary with frequency?
Answer: *The human body absorbs more RF energy at some frequencies than at others.*

The human body resonates more between 35 MHz and 70 MHz than at other frequencies. The Maximum Permissible Exposure limit for this frequency range is shorter for these frequencies than other frequencies.

T0C06 Which of the following is an acceptable method to determine that your station complies with FCC RF exposure regulations?
Answer: *By calculation based on FCC OET Bulletin 65, by calculation based on computer modeling, and by measurement of field strength using calibrated equipment.*

This is another one of those "all of the above" questions. There are several ways of making sure your station is in compliance. You can use the FCC OET Bulletin 65 as a guide. You can make calculations based on computer modeling. Another option is to take measurements of RF field strength around your ham shack and antenna.

T0C07 What could happen if a person accidentally touched your antenna while you were transmitting?
Answer: *They might receive a painful RF burn.*

An RF burn is the result of receiving a solid dose of RF radiation by touching an antenna. The RF radiation heats the skin and results in a burn. An RF burn is painful but not as serious as other types of burns.

T0C08 Which of the following actions might amateur operators take to prevent exposure to RF radiation in excess of FCC-supplied limits?
Answer: *Relocate antennas.*

Of the possible answers, relocating the antenna is the only option that will have any impact on RF radiation exposure. If the antenna is located close to people, causing exposure limits to exceed FCC guidelines, moving the antenna may be the most obvious option.

T0C09 How can you make sure your station stays in compliance with RF safety regulations?
Answer: *By re-evaluating the station whenever an item of equipment is changed.*

The only way to make sure a station stays in compliance with RF safety requirements is to reevaluate when significant changes are made to your equipment. A change in transmitter, amplifier, or antenna can cause an increase in RF radiation exposure. Be sure to reevaluate whenever you change equipment.

T0C10 Why is duty cycle one of the factors used to determine safe RF radiation exposure levels?
Answer: *It affects the average exposure of people to radiation.*

Duty cycle is the ratio of on-air time to total operating time of a transmitted signal. An easier way of saying it is the amount of time a transmitter is actually sending out a signal during a period of time expressed as a fraction. The higher the ratio of transmitting time verses idle time, the higher the potential RF exposure level.

T0C11 What is meant by "duty cycle" when referring to RF exposure?
Answer: *The ratio of on-air time to total operating time of a transmitted signal.*

Duty cycle is the ratio of the time a transmitter is sending a signal compared to the amount of time it is not sending a signal within a specific period. The higher the ratio, the higher the potential RF exposure.

Quiz #35

(Find the answers on page 256)

1. _____ T0C01 What type of radiation are VHF and UHF radio signals?
A. Gamma radiation
B. Ionizing radiation
C. Alpha radiation
D. Non-ionizing radiation

2. _____ T0C02 Which of the following frequencies has the lowest Maximum Permissible Exposure limit?
A. 3.5 MHz
B. 50 MHz
C. 440 MHz
D. 1296 MHz

3. _____ T0C03 What is the maximum power level that an amateur radio station may use at VHF frequencies before an RF exposure evaluation is required?
A. 1500 watts PEP transmitter output
B. 1 watt forward power
C. 50 watts PEP at the antenna
D. 50 watts PEP reflected power

4. _____ T0C04 What factors affect the RF exposure of people near an amateur station antenna?
A. Frequency and power level of the RF field
B. Distance from the antenna to a person
C. Radiation pattern of the antenna
D. All of these choices are correct

5. _____ T0C05 Why do exposure limits vary with frequency?
A. Lower frequency RF fields have more energy than higher frequency fields
B. Lower frequency RF fields do not penetrate the human body
C. Higher frequency RF fields are transient in nature
D. The human body absorbs more RF energy at some frequencies than at others

6. _____ T0C06 Which of the following is an acceptable method to determine that your station complies with FCC RF exposure regulations?
A. By calculation based on FCC OET Bulletin 65
B. By calculation based on computer modeling
C. By measurement of field strength using calibrated equipment
D. All of these choices are correct

7. _____ T0C07 What could happen if a person accidentally touched your antenna while you were transmitting?
A. Touching the antenna could cause television interference
B. They might receive a painful RF burn
C. They might develop radiation poisoning
D. All of these choices are correct

8. _____ T0C08 Which of the following actions might amateur operators take to prevent exposure to RF radiation in excess of FCC-supplied limits?
A. Relocate antennas
B. Relocate the transmitter
C. Increase the duty cycle
D. All of these choices are correct

9. _____ T0C09 How can you make sure your station stays in compliance with RF safety regulations?
A. By informing the FCC of any changes made in your station
B. By re-evaluating the station whenever an item of equipment is changed
C. By making sure your antennas have low SWR
D. All of these choices are correct

10. _____ T0C10 Why is duty cycle one of the factors used to determine safe RF radiation exposure levels?
A. It affects the average exposure of people to radiation
B. It affects the peak exposure of people to radiation
C. It takes into account the antenna feedline loss
D. It takes into account the thermal effects of the final amplifier

11. _____ T0C11 What is meant by "duty cycle" when referring to RF exposure?
A. The difference between lowest usable output and maximum rated output power of a transmitter
B. The difference between PEP and average power of an SSB signal
C. The ratio of on-air time to total operating time of a transmitted signal
D. The amount of time the operator spends transmitting

Lesson 35 Study Sheet

✓ VHF and UHF radio signals are types of non-ionizing radiation.

✓ The lowest Maximum Permissible Exposure limits fall between 35 MHz and 70 MHz.

✓ 50 watts PEP at the antenna is the maximum power level that an amateur radio station may use at VHF frequencies before an RF exposure evaluation is required.

✓ Duty cycle is the ratio of on-air time to total operating time of a transmitted signal.

Notes:

The Easy Exam

Now that you have completed all the lessons, it is time to stretch your mind out and review the material you learned. The purpose behind the three practice exams is to refresh your memory about the subject matter covered in all the lessons and to help identify some problem areas you may want to review before taking the exam. If you think you need some more practice, there are many websites which offer online practice exams. QRZ.com is one of the better practice exam sites. You can find the answers to the three practice exams on page 255.

1. _____T1A10 [97.3(a)(5)] What is the FCC Part 97 definition of an amateur station?
A. A station in an Amateur Radio Service consisting of the apparatus necessary for carrying on radio communications
B. A building where Amateur Radio receivers, transmitters, and RF power amplifiers are installed
C. Any radio station operated by a non-professional
D. Any radio station for hobby use

2. _____T1B09 [97.101(a)] Why should you not set your transmit frequency to be exactly at the edge of an amateur band or sub-band?
A. To allow for calibration error in the transmitter frequency display
B. So that modulation sidebands do not extend beyond the band edge
C. To allow for transmitter frequency drift
D. All of these choices are correct

3. _____T1C08 [97.25] What is the normal term for an FCC-issued primary station/operator license grant?
A. Five years
B. Life
C. Ten years
D. Twenty years

4. _____T1D06 [97.113(a)(4)] Which of the following types of transmissions are prohibited?
A. Transmissions that contain obscene or indecent words or language
B. Transmissions to establish one-way communications
C. Transmissions to establish model aircraft control
D. Transmissions for third party communications

5. _____ T1E02 [97.7(a)] Who is eligible to be the control operator of an amateur station?
A. Only a person holding an amateur service license from any country that belongs to the United Nations
B. Only a citizen of the United States
C. Only a person over the age of 18
D. Only a person for whom an amateur operator/primary station license grant appears in the FCC database or who is authorized for alien reciprocal operation

6. _____ T1F03 [97.119(a)] When is an amateur station required to transmit its assigned call sign?
A. At the beginning of each contact, and every 10 minutes thereafter
B. At least once during each transmission
C. At least every 15 minutes during and at the end of a contact
D. At least every 10 minutes during and at the end of a contact

7. _____T2A08 What is the meaning of the procedural signal "CQ"?
A. Call on the quarter hour
B. A new antenna is being tested (no station should answer)
C. Only the called station should transmit
D. Calling any station

8. _____ T2B01 What is the term used to describe an amateur station that is transmitting and receiving on the same frequency?
A. Full duplex communication
B. Diplex communication
C. Simplex communication
D. Half duplex communication

9. _____ T2C04 What do RACES and ARES have in common?
A. They represent the two largest ham clubs in the United States
B. Both organizations broadcast road and weather traffic information
C. Neither may handle emergency traffic supporting public service agencies
D. Both organizations may provide communications during emergencies

10. _____ T3A07 What type of wave carries radio signals between transmitting and receiving stations?
A. Electromagnetic
B. Electrostatic
C. Surface acoustic
D. Magnetostrictive

11. _____ T3B01 What is the name for the distance a radio wave travels during one complete cycle?
A. Wave speed
B. Waveform
C. Wavelength
D. Wave spread

12. _____ T3C09 What is generally the best time for long-distance 10 meter band propagation?
A. During daylight hours
B. During nighttime hours
C. When there are coronal mass ejections
D. Whenever the solar flux is low

13. _____ T4A02 What could be used in place of a regular speaker to help you copy signals in a noisy area?
A. A video display
B. A low pass filter
C. A set of headphones
D. A boom microphone

14. _____ T4B02 Which of the following can be used to enter the operating frequency on a modern transceiver?
A. The keypad or VFO knob
B. The CTCSS or DTMF encoder
C. The Automatic Frequency Control
D. All of these choices are correct

15. _____T5A07 Which of the following is a good electrical conductor?
A. Glass
B. Wood
C. Copper
D. Rubber

16. _____ T5B03 How many volts are equal to one kilovolt?
A. One one-thousandth of a volt
B. One hundred volts
C. One thousand volts
D. One million volts

17. _____T5C06 What is the abbreviation that refers to radio frequency signals of all types?
A. AF
B. HF
C. RF
D. VHF

18. _____ T5D02 What formula is used to calculate voltage in a circuit?
A. Voltage (E) equals current (I) multiplied by resistance (R)
B. Voltage (E) equals current (I) divided by resistance (R)
C. Voltage (E) equals current (I) added to resistance (R)
D. Voltage (E) equals current (I) minus resistance (R)

19. _____ T6A05 What type of electrical component consists of two or more conductive surfaces separated by an insulator?
A. Resistor
B. Potentiometer
C. Oscillator
D. Capacitor

20. _____ T6B07 What does the abbreviation "LED" stand for?
A. Low Emission Diode
B. Light Emitting Diode
C. Liquid Emission Detector
D. Long Echo Delay

21. _____ T6C12 What do the symbols on an electrical circuit schematic diagram represent?
A. Electrical components
B. Logic states
C. Digital codes
D. Traffic nodes

22. _____ T6D04 Which of the following can be used to display signal strength on a numeric scale?
A. Potentiometer
B. Transistor
C. Meter
D. Relay

23. _____ T7A10 What device increases the low-power output from a handheld transceiver?
A. A voltage divider
B. An RF power amplifier
C. An impedance network
D. A voltage regulator

24. _____ T7B10 What might be the problem if you receive a report that your audio signal through the repeater is distorted or unintelligible?
A. Your transmitter may be slightly off frequency
B. Your batteries may be running low
C. You could be in a bad location
D. All of these choices are correct

25. _____ T7C04 What reading on an SWR meter indicates a perfect impedance match between the antenna and the feedline?
A. 2 to 1
B. 1 to 3
C. 1 to 1
D. 10 to 1

26. _____ T7D04 Which instrument is used to measure electric current?
A. An ohmmeter
B. A wavemeter
C. A voltmeter
D. An ammeter

27. _____ T8A02 What type of modulation is most commonly used for VHF packet radio transmissions?
A. FM
B. SSB
C. AM
D. Spread Spectrum

28. _____ T8B02 [97.313(a)] How much transmitter power should be used on the uplink frequency of an amateur satellite or space station?
A. The maximum power of your transmitter
B. The minimum amount of power needed to complete the contact
C. No more than half the rating of your linear amplifier
D. Never more than 1 watt

29. _____ T8C03 What popular operating activity involves contacting as many stations as possible during a specified period of time?
A. Contesting
B. Net operations
C. Public service events
D. Simulated emergency exercises

30. _____ T8D09 What code is used when sending CW in the amateur bands?
A. Baudot
B. Hamming
C. International Morse
D. Gray

31. _____ T9A06 What type of antennas are the quad, Yagi, and dish?
A. Non-resonant antennas
B. Loop antennas
C. Directional antennas
D. Isotropic antennas

32. _____ T9B03 Why is coaxial cable used more often than any other feedline for amateur radio antenna systems?
A. It is easy to use and requires few special installation considerations
B. It has less loss than any other type of feedline
C. It can handle more power than any other type of feedline
D. It is less expensive than any other types of feedline

33. _____ T0A02 How does current flowing through the body cause a health hazard?
A. By heating tissue
B. It disrupts the electrical functions of cells
C. It causes involuntary muscle contractions
D. All of these choices are correct

34. _____ T0B03 Under what circumstances is it safe to climb a tower without a helper or observer?
A. When no electrical work is being performed
B. When no mechanical work is being performed
C. When the work being done is not more than 20 feet above the ground
D. Never

35. _____ T0C07 What could happen if a person accidentally touched your antenna while you were transmitting?
A. Touching the antenna could cause television interference
B. They might receive a painful RF burn
C. They might develop radiation poisoning
D. All of these choices are correct

Challenging Exam

1. _____ T1A03 Which part of the FCC rules contains the rules and regulations governing the Amateur Radio Service?
A. Part 73
B. Part 95
C. Part 90
D. Part 97

2. _____ T1B11 [97.305 (a)(c)] What emission modes are permitted in the mode-restricted sub-bands at 50.0 to 50.1 MHz and 144.0 to 144.1 MHz?
A. CW only
B. CW and RTTY
C. SSB only
D. CW and SSB

3. _____ T1C07 [97.23] What may result when correspondence from the FCC is returned as undeliverable because the grantee failed to provide the correct mailing address?
A. Fine or imprisonment
B. Revocation of the station license or suspension of the operator license
C. Require the licensee to be re-examined
D. A reduction of one rank in operator class

4. _____ T1D11 [97.113(a)(5)] Which of the following types of communications are permitted in the Amateur Radio Service?
A. Brief transmissions to make station adjustments
B. Retransmission of entertainment programming from a commercial radio or TV station
C. Retransmission of entertainment material from a public radio or TV station
D. Communications on a regular basis that could reasonably be furnished alternatively through other radio services

5. _____ T1E01 [97.7(a)] When must an amateur station have a control operator?
A. Only when the station is transmitting
B. Only when the station is being locally controlled
C. Only when the station is being remotely controlled
D. Only when the station is being automatically controlled

6. _____ T1F05 [97.119(b)] What method of call sign identification is required for a station transmitting phone signals?
A. Send the call sign followed by the indicator RPT
B. Send the call sign using CW or phone emission
C. Send the call sign followed by the indicator R
D. Send the call sign using only phone emission

7. _____ T2A06 What must an amateur operator do when making on-air transmissions to test equipment or antennas?
A. Properly identify the transmitting station
B. Make test transmissions only after 10:00 p.m. local time
C. Notify the FCC of the test transmission
D. State the purpose of the test during the test procedure

8. _____ T2B07 What should you do if you receive a report that your station's transmissions are causing splatter or interference on nearby frequencies?
A. Increase transmit power
B. Change mode of transmission
C. Report the interference to the equipment manufacturer
D. Check your transmitter for off-frequency operation or spurious emission

9. _____ T2C05 [97.3(a)(37), 97.407] What is the Radio Amateur Civil Emergency Service?
A. An emergency radio service organized by amateur operators
B. A radio service using amateur stations for emergency management or civil defense communications
C. A radio service organized to provide communications at civic events
D. A radio service organized by amateur operators to assist non-military persons

10. _____ T3A02 Why are UHF signals often more effective from inside buildings than VHF signals?
A. VHF signals lose power faster over distance
B. The shorter wavelength allows them to more easily penetrate the structure of buildings
C. This is incorrect; VHF works better than UHF inside buildings
D. UHF antennas are more efficient than VHF antennas

11. _____ T3B05 How does the wavelength of a radio wave relate to its frequency?
A. The wavelength gets longer as the frequency increases
B. The wavelength gets shorter as the frequency increases
C. There is no relationship between wavelength and frequency
D. The wavelength depends on the bandwidth of the signal

12. _____ T3C07 What band is best suited to communicating via meteor scatter?
A. 10 meters
B. 6 meters
C. 2 meters
D. 70 cm

13. _____ T4A06 Which of the following would be connected between a transceiver and computer in a packet radio station?
A. Transmatch
B. Mixer
C. Terminal node controller
D. Antenna

14. _____ T4B07 What does the term "RIT" mean?
A. Receiver Input Tone
B. Receiver Incremental Tuning
C. Rectifier Inverter Test
D. Remote Input Transmitter

15. _____ T5A05 What is the electrical term for the electromotive force (EMF) that causes electron flow?
A. Voltage
B. Ampere-hours
C. Capacitance
D. Inductance

16. _____ T5B02 What is another way to specify a radio signal frequency of 1,500,000 hertz?
A. 1500 kHz
B. 1500 MHz
C. 15 GHz
D. 150 kHz

17. _____ T5C03 What is the ability to store energy in a magnetic field called?
A. Admittance
B. Capacitance
C. Resistance
D. Inductance

18. _____ T5D10 What is the voltage across a 2-ohm resistor if a current of 0.5 amperes flows through it?
A. 1 volt
B. 0.25 volts
C. 2.5 volts
D. 1.5 volts

19. _____ T6A09 What electrical component is used to protect other circuit components from current overloads?
A. Fuse
B. Capacitor
C. Shield
D. Inductor

20. _____ T6B01 What class of electronic components is capable of using a voltage or current signal to control current flow?
A. Capacitors
B. Inductors
C. Resistors
D. Transistors

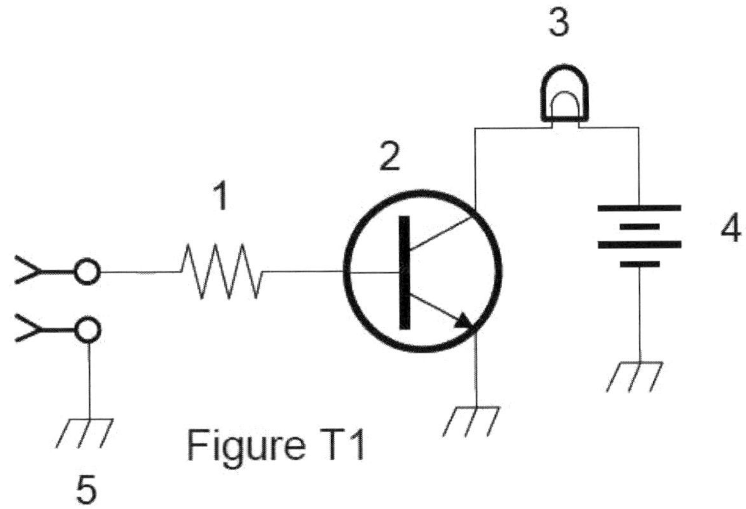

Figure T1

21. _____ T6C04 What is component 3 in figure T1?
A. Resistor
B. Transistor
C. Lamp
D. Ground symbol

22. _____ T6D02 What best describes a relay?
A. A switch controlled by an electromagnet
B. A current controlled amplifier
C. An optical sensor
D. A pass transistor

23. _____ T7A09 Which of the following devices is most useful for VHF weak-signal communication?
A. A quarter-wave vertical antenna
B. A multi-mode VHF transceiver
C. An omni-directional antenna
D. A mobile VHF FM transceiver

24. _____ T7B09 What could be happening if another operator reports a variable high-pitched whine on the audio from your mobile transmitter?
A. Your microphone is picking up noise from an open window
B. You have the volume on your receiver set too high
C. You need to adjust your squelch control
D. Noise on the vehicle's electrical system is being transmitted along with your speech audio

25. _____ T7C02 Which of the following instruments can be used to determine if an antenna is resonant at the desired operating frequency?
A. A VTVM
B. An antenna analyzer
C. A "Q" meter
D. A frequency counter

26. _____ T7D08 Which of the following types of solder is best for radio and electronic use?
A. Acid-core solder
B. Silver solder
C. Rosin-core solder
D. Aluminum solder

27. _____ T8A05 Which of the following types of emission has the narrowest bandwidth?
A. FM voice
B. SSB voice
C. CW
D. Slow-scan TV

28. _____ T8B04 Which amateur stations may make contact with an amateur station on the International Space Station using 2 meter and 70 cm band amateur radio frequencies?
A. Only members of amateur radio clubs at NASA facilities
B. Any amateur holding a Technician or higher class license
C. Only the astronaut's family members who are hams
D. You cannot talk to the ISS on amateur radio frequencies

29. _____ T8C02 Which of these items would be useful for a hidden transmitter hunt?
A. Calibrated SWR meter
B. A directional antenna
C. A calibrated noise bridge
D. All of these choices are correct

30. _____ T8D06 What does the abbreviation PSK mean?
A. Pulse Shift Keying
B. Phase Shift Keying
C. Packet Short Keying
D. Phased Slide Keying

31. _____ T9A03 Which of the following describes a simple dipole mounted so the conductor is parallel to the Earth's surface?
A. A ground wave antenna
B. A horizontally polarized antenna
C. A rhombic antenna
D. A vertically polarized antenna

32. _____ T9B09 What might cause erratic changes in SWR readings?
A. The transmitter is being modulated
B. A loose connection in an antenna or a feedline
C. The transmitter is being over-modulated
D. Interference from other stations is distorting your signal

33. _____ T0A03 What is connected to the green wire in a three-wire electrical AC plug?
A. Neutral
B. Hot
C. Safety ground
D. The white wire

34. _____ T0B06 What is the minimum safe distance from a power line to allow when installing an antenna?
A. Half the width of your property
B. The height of the power line above ground
C. 1/2 wavelength at the operating frequency
D. So that if the antenna falls unexpectedly, no part of it can come closer than 10 feet to the power wires

35. _____ T0C06 Which of the following is an acceptable method to determine that your station complies with FCC RF exposure regulations?
A. By calculation based on FCC OET Bulletin 65
B. By calculation based on computer modeling
C. By measurement of field strength using calibrated equipment
D. All of these choices are correct

Hard Exam

1. _____ T1A07 [97.3(a)(45)] What is the FCC Part 97 definition of telemetry?
A. An information bulletin issued by the FCC
B. A one-way transmission to initiate, modify or terminate functions of a device at a distance
C. A one-way transmission of measurements at a distance from the measuring instrument
D. An information bulletin from a VEC

2. _____ T1B05 [97.301(a)] Which 70 cm frequency is authorized to a Technician Class license holder operating in ITU Region 2?
A. 53.350 MHz
B. 146.520 MHz
C. 443.350 MHz
D. 222.520 MHz

3. _____ T1C06 [97.5(a)(2)] From which of the following may an FCC-licensed amateur station transmit, in addition to places where the FCC regulates communications?
A. From within any country that belongs to the International Telecommunications Union
B. From within any country that is a member of the United Nations
C. From anywhere within in ITU Regions 2 and 3
D. From any vessel or craft located in international waters and documented or registered in the United States

4. _____ T1D01 [97.111(a)(1)] With which countries are FCC-licensed amateur stations prohibited from exchanging communications?
A. Any country whose administration has notified the ITU that it objects to such communications
B. Any country whose administration has notified the United Nations that it objects to such communications
C. Any country engaged in hostilities with another country
D. Any country in violation of the War Powers Act of 1934

5. _____ T1E10 [97.3] What type of control is used when the control operator is not at the station location but can indirectly manipulate the operating adjustments of a station?
A. Local
B. Remote
C. Automatic
D. Unattended

6. _____ T1F06 [97.119(c)] Which of the following formats of a self-assigned indicator is acceptable when identifying using a phone transmission?
A. KL7CC stroke W3
B. KL7CC slant W3
C. KL7CC slash W3
D. All of these choices are correct

7. _____ T2A03 What is a common repeater frequency offset in the 70 cm band?
A. Plus or minus 5 MHz
B. Plus or minus 600 kHz
C. Minus 600 kHz
D. Plus 600 kHz

8. _____ T2B02 What is the term used to describe the use of a sub-audible tone transmitted with normal voice audio to open the squelch of a receiver?
A. Carrier squelch
B. Tone burst
C. DTMF
D. CTCSS

9. _____ T2C03 [97.113] When is it legal for an amateur licensee to provide communications on behalf of their employer during a government sponsored disaster drill or exercise?
A. Whenever the employer is a not-for-profit organization
B. Whenever there is a temporary need for the employer's business continuity plan
C. Only when the FCC has granted a government-requested waiver
D. Only when the amateur is not receiving compensation from his employer for the activity

10. _____ T3A08 What is the cause of irregular fading of signals from distant stations during times of generally good reception?
A. Absorption of signals by the "D" layer of the ionosphere
B. Absorption of signals by the "E" layer of the ionosphere
C. Random combining of signals arriving via different path lengths
D. Intermodulation distortion in the local receiver

11. _____ T3B06 What is the formula for converting frequency to wavelength in meters?
A. Wavelength in meters equals frequency in hertz multiplied by 300
B. Wavelength in meters equals frequency in hertz divided by 300
C. Wavelength in meters equals frequency in megahertz divided by 300
D. Wavelength in meters equals 300 divided by frequency in megahertz

12. _____ T3C05 What is meant by the term "knife-edge" propagation?
A. Signals are reflected back toward the originating station at acute angles
B. Signals are sliced into several discrete beams and arrive via different paths
C. Signals are partially refracted around solid objects exhibiting sharp edges
D. Signals propagated close to the band edge exhibiting a sharp cutoff

13. _____ T4A05 What type of filter should be connected to a TV receiver as the first step in trying to prevent RF overload from a nearby 2 meter transmitter?
A. Low-pass filter
B. High-pass filter
C. Band-pass filter
D. Band-reject filter

14. _____ T4B09 Which of the following is an appropriate receive filter to select in order to minimize noise and interference for SSB reception?
A. 500 Hz
B. 1000 Hz
C. 2400 Hz
D. 5000 Hz

15. _____ T5A09 What is the name for a current that reverses direction on a regular basis?
A. Alternating current
B. Direct current
C. Circular current
D. Vertical current

16. _____ T5B08 How many microfarads are 1,000,000 picofarads?
A. 0.001 microfarads
B. 1 microfarad
C. 1000 microfarads
D. 1,000,000,000 microfarads

17. _____ T5C09 How much power is being used in a circuit when the applied voltage is 13.8 volts DC and the current is 10 amperes?
A. 138 watts
B. 0.7 watts
C. 23.8 watts
D. 3.8 watts

18. _____ T5D09 What is the current flowing through a 24-ohm resistor connected across 240 volts?
A. 24,000 amperes
B. 0.1 amperes
C. 10 amperes
D. 216 amperes

19. _____ T6A02 What type of component is often used as an adjustable volume control?
A. Fixed resistor
B. Power resistor
C. Potentiometer
D. Transformer

20. _____ T6B03 Which of these components can be used as an electronic switch or amplifier?
A. Oscillator
B. Potentiometer
C. Transistor
D. Voltmeter

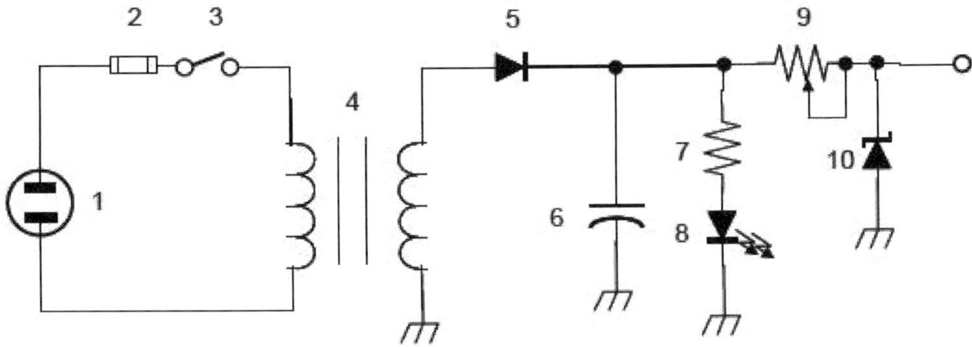

Figure T2

21. _____ T6C09 What is component 4 in figure T2?
A. Variable inductor
B. Double-pole switch
C. Potentiometer
D. Transformer

22. _____ T6D01 Which of the following devices or circuits changes an alternating current into a varying direct current signal?
A. Transformer
B. Rectifier
C. Amplifier
D. Reflector

23. _____ T7A03 What is the function of a mixer in a superheterodyne receiver?
A. To reject signals outside of the desired passband
B. To combine signals from several stations together
C. To shift the incoming signal to an intermediate frequency
D. To connect the receiver with an auxiliary device, such as a TNC

24. _____ T7B12 What does the acronym "BER" mean when applied to digital communications systems?
A. Baud Enhancement Recovery
B. Baud Error Removal
C. Bit Error Rate
D. Bit Exponent Resource

25. _____ T7C05 What is the approximate SWR value above which the protection circuits in most solid-state transmitters begin to reduce transmitter power?
A. 2 to 1
B. 1 to 2
C. 6 to 1
D. 10 to 1

26. _____ T7D06 Which of the following might damage a multimeter?
A. Measuring a voltage too small for the chosen scale
B. Leaving the meter in the milliamps position overnight
C. Attempting to measure voltage when using the resistance setting
D. Not allowing it to warm up properly

27. _____ T8A09 What is the approximate bandwidth of a VHF repeater FM phone signal?
A. Less than 500 Hz
B. About 150 kHz
C. Between 5 and 15 kHz
D. Between 50 and 125 kHz

28. _____ T8B07 With regard to satellite communications, what is Doppler shift?
A. A change in the satellite orbit
B. A mode where the satellite receives signals on one band and transmits on another
C. An observed change in signal frequency caused by relative motion between the satellite and the earth station
D. A special digital communications mode for some satellites

29. _____ T8C10 How do you select a specific IRLP node when using a portable transceiver?
A. Choose a specific CTCSS tone
B. Choose the correct DSC tone
C. Access the repeater autopatch
D. Use the keypad to transmit the IRLP node ID

30. _____ T8D04 What type of transmission is indicated by the term NTSC?
A. A Normal Transmission mode in Static Circuit
B. A special mode for earth satellite uplink
C. An analog fast scan color TV signal
D. A frame compression scheme for TV signals

31. _____ T9A08 What is the approximate length, in inches, of a quarter-wavelength vertical antenna for 146 MHz?
A. 112
B. 50
C. 19
D. 12

32. _____ T9B05 What generally happens as the frequency of a signal passing through coaxial cable is increased?
A. The apparent SWR increases
B. The reflected power increases
C. The characteristic impedance increases
D. The loss increases

33. _____ T0A07 Which of these precautions should be taken when installing devices for lightning protection in a coaxial cable feedline?
A. Include a parallel bypass switch for each protector so that it can be switched out of the circuit when running high power
B. Include a series switch in the ground line of each protector to prevent RF overload from inadvertently damaging the protector
C. Keep the ground wires from each protector separate and connected to station ground
D. Ground all of the protectors to a common plate which is in turn connected to an external ground

34. _____ T0B05 What is the purpose of a gin pole?
A. To temporarily replace guy wires
B. To be used in place of a safety harness
C. To lift tower sections or antennas
D. To provide a temporary ground

35. _____ T0C02 Which of the following frequencies has the lowest Maximum Permissible Exposure limit?
A. 3.5 MHz
B. 50 MHz
C. 440 MHz
D. 1296 MHz

Where To Go From Here

The world can seem like a lonely place for the newly licensed Technician Class amateur operator. Getting on the air for the first time can be an intimidating procedure when the only real ham radio experience you have is the fact you passed the exam.

There are several bits of advice I received that helped me out when I received my first license. The first is to just sit back and listen to the receiver and become familiar with the way hams talk. Take a few hours and just listen to what's going on. Take notes if you'd like. You learn an awful lot just by listening to other amateurs make contacts.

The second piece of advice I received was to join a local ham radio club. Tune in the local 2 meter repeater and listen for a club net. Nets are on air meetings of groups of hams. If a local club sponsors the repeater, chances are that club has a regularly scheduled net on that repeater where they pass announcements and take a roll call. Listen to the net and find out some information about the club. Attend a few meetings and get to know the local hams in your area. When your operating skills are still a little green, being able to put a face with the call sign helps build some confidence.

One of the best pieces of advice is to join the American Radio Relay League (ARRL) as soon as possible! Membership definitely does have its privileges! Check out the website at www.ARRL.org. There are thousands of pages of information for amateur radio operators and even a section for new hams! The ARRL publishes dozens of books as well as their monthly magazine, QST.

Now all you need to do is push the transmit button and get on the air! A big worry many new hams have is they are nervous about sounding like a new ham when they get on the air. Don't worry about it! In fact, advertise the fact that you just received your license! You will be quickly bombarded with a wealth of advice and encouragement. The fact is the only way to build operating skill is to get on the air! It's a thrilling experience! Enjoy it!

For additional information, helpful links, and video tutorials, stop by www.HamWhisperer.com.

Practice Exam Answers

Easy Exam			Challenging Exam			Hard Exam		
Q	A	Question Pool Number	Q	A	Question Pool Number	Q	A	Question Pool Number
1.	A	T1A10	1.	D	T1A03	1.	C	T1A07
2.	D	T1B09	2.	A	T1B11	2.	C	T1B05
3.	C	T1C08	3.	B	T1C07	3.	D	T1C06
4.	A	T1D06	4.	A	T1D11	4.	A	T1D01
5.	C	T1E02	5.	A	T1E01	5.	B	T1E10
6.	D	T1F03	6.	B	T1F05	6.	D	T1F06
7.	D	T2A08	7.	A	T2A06	7.	A	T2A03
8.	C	T2B01	8.	D	T2B07	8.	D	T2B02
9.	D	T2C04	9.	B	T2C05	9.	C	T2C03
10.	A	T3A07	10.	B	T3A02	10.	C	T3A08
11.	C	T3B01	11.	B	T3B05	11.	D	T3B06
12.	A	T3C09	12.	B	T3C07	12.	C	T3C05
13.	C	T4A02	13.	C	T4A06	13.	D	T4A05
14.	A	T4B02	14.	B	T4B07	14.	C	T4B09
15.	C	T5A07	15.	A	T5A05	15.	A	T5A09
16.	C	T5B03	16.	A	T5B02	16.	B	T5B08
17.	C	T5C06	17.	D	T5C03	17.	A	T5C09
18.	A	T5D02	18.	A	T5D10	18.	C	T5D09
19.	D	T6A05	19.	A	T6A09	19.	C	T6A02
20.	B	T6B07	20.	D	T6B01	20.	C	T6B03
21.	A	T6C12	21.	A	T6C02	21.	D	T6C09
22.	C	T6D04	22.	A	T6D02	22.	B	T6D01
23.	B	T7A10	23.	B	T7A09	23.	C	T7A03
24.	D	T7B10	24.	D	T7B09	24.	C	T7B12
25.	C	T7C04	25.	B	T7C02	25.	A	T7C05
26.	D	T7D04	26.	C	T7D08	26.	C	T7D06
27.	A	T8A02	27.	C	T8A05	27.	C	T8A09
28.	B	T8B02	28.	B	T8B04	28.	C	T8B07
29.	A	T8C03	29.	B	T8C02	29.	D	T8C10
30.	C	T8D09	30.	B	T8D06	30.	C	T8D04
31.	C	T9A06	31.	B	T9A03	31.	C	T9A08
32.	A	T9B03	32.	B	T9B09	32.	D	T9B05
33.	D	T0A02	33.	C	T0A03	33.	D	T0A07
34.	D	T0B03	34.	D	T0B06	34.	C	TOB05
35.	B	T0C07	35.	D	T0C06	35.	B	T0C02

Quiz Answers

Quiz #												
#1	1. D	2. C	3. D	4. C	5. D	6. C	7. C	8. B	9. C	10. A	11. C		
#2	1. B	2. B	3. B	4. A	5. C	6. B	7. D	8. C	9. D	10. C	11. A		
#3	1. C	2. B	3. A	4. A	5. A	6. D	7. B	8. C	9. A	10. C	11. A		
#4	1. A	2. A	3. C	4. A	5. A	6. A	7. B	8. B	9. A	10. D	11. A		
#5	1. A	2. D	3. A	4. D	5. C	6. B	7. D	8. C	9. D	10. B	11. D		
#6	1. A	2. C	3. D	4. C	5. B	6. D	7. D	8. A	9. C	10. A	11. A	12. B	13. B
#7	1. B	2. D	3. A	4. B	5. C	6. A	7. D	8. D	9. B	10. A	11. D		
#8	1. C	2. D	3. B	4. D	5. C	6. A	7. D	8. B	9. A	10. A	11. B		
#9	1. C	2. D	3. C	4. D	5. B	6. C	7. C	8. A	9. B	10. D	11. A		
#10	1.D	2. B	3. C	4. B	5. B	6. B	7. A	8. C	9. B	10. D	11. C		
#11	1. C	2. D	3. C	4. A	5. B	6. D	7. A	8. B	9. D	10. C	11. B		
#12	1. C	2. D	3. B	4. B	5. C	6. A	7. B	8. D	9. A	10. A	11. C		
#13	1. B	2. C	3. A	4. A	5. D	6. C	7. C	8. D	9. D	10. B	11. A		
#14	1. B	2. A	3. D	4. B	5. C	6. D	7. B	8. B	9. C	10. A	11. C		
#15	1. D	2.B	3. D	4. B	5. A	6. A	7. C	8. B	9. A	10. C	11. A		
#16	1. C	2. A	3. C	4. A	5. B	6. C	7. C	8. B	9. B	10. C	11. A		
#17	1. D	2. A	3. D	4. C	5. A	6. C	7. C	8. A	9. A	10. B	11. B		
#18	1. B	2. A	3. B	4. B	5. C	6. A	7. D	8. C	9. C	10. A	11. B	12. D	
#19	1. B	2. C	3. B	4. B	5. D	6. C	7. D	8. B	9. A	10. B	11. B		
#20	1. D	2. C	3. C	4. B	5. A	6. B	7. B	8. A	9. C	10. A	11. B	12. A	
#21	1. C	2. A	3. B	4. C	5. C	6. B	7. D	8. C	9. D	10. D	11. A	12. A	13. C
#22	1. B	2. A	3. A	4. C	5. A	6. B	7. A	8. D	9. C	10. C	11. B		
#23	1. C	2. C	3. C	4. D	5. D	6. C	7. B	8. C	9. B	10. B	11. B	12. C	13. A
#24	1. D	2. C	3. D	4. B	5. C	6. A	7. D	8. D	9. D	10. D	11. C	12. C	
#25	1. A	2. B	3. A	4. C	5. A	6. D	7. C	8. D	9. A	10. D	11. C		
#26	1. B	2. B	3. A	4. D	5. D	6. C	7. D	8. C	9. C	10. B	11. B		
#27	1. C	2. A	3. C	4. D	5. C	6. A	7. C	8. B	9. C	10. B	11. B		
#28	1. D	2. B	3. A	4. B	5. D	6. D	7. C	8. B	9. B	10. C	11. C		
#29	1. C	2. B	3. A	4. C	5. A	6. C	7. B	8. C	9. C	10. D	11. A		
#30	1. D	2. A	3. D	4. C	5. B	6. B	7. D	8. D	9. C	10. D	11. C		
#31	1. C	2. B	3. B	4. A	5. C	6. C	7. A	8. C	9. C	10. C	11. C		
#32	1. B	2. B	3. A	4. A	5. D	6. B	7. C	8. A	9. B	10. C	11. C		
#33	1. B	2. D	3. C	4. B	5. C	6. D	7. D	8. C	9. C	10. A	11. C	12. D	13. A
#34	1. C	2. C	3. D	4. C	5. C	6. D	7. C	8. C	9. C	10. C	11. B		
#35	1. D	2. B	3. C	4. D	5. D	6. D	7. B	8. A	9. B	10. A	11. C		

Made in the USA
San Bernardino, CA
03 March 2014